Genesis of Fighter Aviation in WWI

A Centennial Perspective on Great War Aviation

Jack Herris

Great War Aviation Centennial Series #20

This book is dedicated to the fighter pilots of the Great War, who invented new ways to live, new ways to fight, and new ways to die, thereby inspiring us with their courage, skill, and sacrifice.

Acknowledgements

My sincere thanks to Greg VanWyngarden for photographs and helpful suggestions, Colin Owers for his photos, drawings, and helpful advice, Lance Bronnenkant, Richard Andrews, Dave Hooper, and Bill Toohey for photographs, Reinhard Zankl for photos and technical information, Bob Pearson for his color profiles, Aaron Weaver for photographs, image editing, and his cover design, Steve Anderson for his cover painting, and the Museum of Flight in Seattle and the Deutsches Technikmuseum in Berlin for photographs. Any errors are my responsibility.

Cover painting *Flying Hussar* by Steve Anderson. Please see his website: **www.anderson-art.com**

Color aircraft profiles © Bob Pearson. Purchase his CD of WWI aircraft profiles for $50 US/Canadian, 40 €, or £30, airmail postage included, via Paypal to Bob at: **bpearson@kaien.net**

For our aviation books, please see our website at: **www.aeronautbooks.com**.

I am looking for photographs of the less well-known German aircraft of WWI to complete this series. For questions or to help with photographs you may contact me at **jherris@me.com**

ISBN: **978-1-935881-36-0**

Design and layout: Jack Herris
Cover design: Aaron Weaver
Digital photo editing: Aaron Weaver & Jack Herris

Aeronaut Books

www.aeronautbooks.com

Table of Contents

WWI Aviation Production

Country	Airplanes	Aero Engines
Austria-Hungary	5,286	4,902
Germany	50,296	43,486
Total Central Powers	55,582	48,388
France	51,700	92,386
Italy	11,986	14,849
Russia	5,607	1,511
United Kingdom	52,027	41,034
United States	10,980	32,420
Total Allies	132,300	182,200

Forward

This is the story of the origin and development of fighter aviation during World War I, the first air war, and the legend of the fighter ace. The story combines the exploits of the aces, the evolution of combat tactics, and the technical development of fighter aircraft from its primitive beginnings to the armistice, presaging the aerial arms race that continues today, when the fighter is the key offensive aerial weapon.

To help make sense of technical development we will make use of the concept of aircraft generations, a concept that has been used for a long time to discuss technical evolution of aircraft, and Richard Hallion used it effectively to discuss WWI fighters in his fine book *Rise of the Fighter Aircraft 1914–1918*. However, we will attempt to clarify the factors used to assign a specific aircraft design to a particular generation, and this book differs from Hallion about the number of fighter generations and to which generation some WWI fighters belong. Of course, the reader's opinion may differ from both of us, but discussion of these differences may prove enlightening.

WWI Fighter Generations

Generation	Attributes
0	Pre-war design, before fighters
1st Generation	Pre-war airframe, armed
2nd Generation	Designed as a fighter during war, but without benefit of operational experience for guidance
3rd Generation	First generation of fighters designed with operational experience for guidance
4th Generation	Second generation of fighters designed with operational experience for guidance
5th Generation	Fighters with advanced engines or advanced airframes & structures
6th Generation	Fighters of all metal construction

Above: WW1 fighter generations and their attributes as used in this book.

Left: This table shows the huge numerical advantage the Allies enjoyed in aircraft and aero-engine production. The limiting issue was engine production; engines are much harder to produce than airframes because they require much tighter production tolerances, more sophisticated tooling, and more highly-skilled workers. All Allied countries depended on France for aviation technology, aircraft, and aero-engines. France's immense aero-engine production far surpassed that of any other country and was essential to the Allied numerical advantage in the air.

Introduction

Throughout WWI the most important role of aircraft was reconnaissance. For the first time in history, aircraft enabled combatants to perform tactical and strategic reconnaissance over one another's military forces from the air. For the first time, armies had to maneuver in full view of their opponent's eyes and cameras. This new reality changed warfare forever and made aviation critical to success.

While aviation explored many roles during the war, the modest performance and payload of WWI airplanes limited their effectiveness when bombing and strafing. These offensive roles became more important as tactics and technology advanced during the war, but were seldom decisive in combat as they became in WW2 and later conflicts. During WWI reconnaissance remained aviation's key contribution to victory.

Types of Aerial Reconnaissance

Aerial reconnaissance included visual reconnaissance, typically with the observer taking notes and making drawings, photographic reconnaissance, and artillery spotting. All these functions could be done over short ranges by tethered observation balloons, but airplanes could see farther behind the enemy lines with a better angle of view, so were generally preferred for these tasks.

Although WWI aircraft lacked the ability to carry large bomb loads, an artillery spotting aircraft could observe the artillery's fall of shot and send back aiming corrections via signals, dropping messages, or wireless. Wireless, which used telegraphy at the time, not voice due to technology limitations, was the fastest and preferred method for giving feedback to artillery batteries. In this way, aircraft were critical to effective artillery fire when the artillery was trying to destroy specific targets. The fairly short range of artillery meant the spotting aircraft had to fly directly over the enemy's front lines, attracting anti-aircraft fire and enemy fighters like a magnet.

Fighter aviation developed to keep the enemy's reconnaissance aircraft from effectively performing their duties. If the most important role of aviation was reconnaissance, the most important role of fighter aviation was to defeat the enemy's aerial reconnaissance efforts.

Technology

The Wright brother's first powered flight was on December 17, 1903, and WWI started in August 1914, less than eleven years later. Aviation technology was in its infancy at the start of the war and made few fundamental breakthroughs during the conflict, although great strides were made in performance and reliability.

The Combatants

The combatants combined in two opposing groups. The Central Powers included Germany, the Austro-Hungarian Empire, and soon the Ottoman Empire and Bulgaria. The Entende, or Allied Powers, included Great Britain, France, and Russia. In 1915 Italy joined the Allies and the United States joined the Allies in April 1917.

The Central Powers were predominantly land powers. Most of the Allied powers were also predominantly land powers, with the notable exception of Great Britain, which was by far the world's largest sea power.

Once the war became a war of attrition, as shown by the transition to trench warfare on the Western Front in October 1914, Great Britain's dominance at sea became decisive. The Royal Navy adopted the strategy of distant blockade, which cut the Central Powers off from overseas resources of men and material. The Royal Navy also ensured the Allies had access to the world's material resources and the immense manpower reserves of the British Empire and the French overseas colonies.

Although there was intense fighting on land and in the air, the war at sea was generally quiescent except for the Allied war against German submarines. Nevertheless, British dominance at sea eventually determined the war's outcome in the Allies' favor. Germany was starved of resources of all kinds, including food, and German forces on land, in the air, and at sea were always out-numbered by Allied forces except in local, tactical situations.

The Allied superiority in resources drove the Allied and Germany strategies, tactics, and procurement policies. The Allies worked to maximize their numerical advantage by producing large numbers of successful airplane designs, while the Germans developed a greater variety of aircraft for different roles to maximize their effectiveness despite inferiority in numbers. The Allies also had resources to invest in many different aero-engine designs to achieve superior performance, while Germany focused on maximizing production of standardized engines and designed different airframes to maximize combat effectiveness.

Before the Fighter

Before WWI there had not been any actual air-to-air combat. In the resulting vacuum of experience, theory naturally prevailed. The most wide-spread theory was based on previous experience, and the most relevant precedent was naval warfare. Thus, many pre-war combat aircraft were designed as 'aerial cruisers', with the pilot flying the aircraft and one or more gunners firing flexible machine guns at the enemy. These aircraft were often expected to carry bombs as well, being all-purpose 'battleplanes', foreshadowing later 'multi-role aircraft.'

Both sides designed and built aircraft to the 'battle plane' or 'aerial cruiser' concept. Experience soon proved these aircraft were too slow to intercept enemy aircraft and had limited effectiveness in air-to-air combat. Tactics consisted of flying close to an enemy aircraft and letting the gunner(s) fire at it. However, if the enemy aircraft had a gunner with flexible gun, the odds were basically even.

Cannon were tried but were heavy, single-shot weapons that had to be reloaded by hand, a challenging process during air-to-air combat. Therefore most attention was directed to the flexible machine gun, which was much lighter and could fire repeatedly, a key attribute when attacking a moving target.

Allied 'battleplanes' included types like Britain's Vickers F.B.5 'Gun Bus'. The F.B.5 featured a typical configuration for the time, a two-seat biplane pusher design with the gunner in front to maximize his field of fire. The lack of effective synchronizing gear meant that a tractor configuration impeded the ability of a crewman to fire forward, a serious limitation. The French Deperdussin tractor

The 'Scout'

During WWI fighter aircraft were called 'scouts'. The true scout aircraft was a small, relatively fast (for the time), unarmed, one or two-seat airplane designed for visual reconnaissance, or scouting. The first true fighters were developed from these scout aircraft by arming them with machine guns, and the name 'scout' stuck for the remainder of WWI. Part of the reason was the original, unarmed scout aircraft designed for visual reconnaissance quickly became obsolete. The reconnaissance role was taken over by two-seat aircraft, which soon replaced the single-seat 'scouts' for visual reconnaissance and presently started carrying cameras for photographic reconnaissance. Interestingly, armed fighters were often used for visual reconnaissance and a few were even fitted with cameras for photographic reconnaissance. However, two-seaters were the primary reconnaissance aircraft throughout the war.

Left: The Breguet BLC was a French pusher two-seat fighter. As late as February 1916 this aircraft, *Vérité*, was being flown as an escort fighter to bombers of *GB* 4 by *Capitaine* Gaubert of *Detachment de Chasse Breguet* at Belfort. The pilot sat in back and the gunner sat in the front cockpit with a flexible machine gun.

Above: The Deperdussin TT could be armed with a flexible machine gun able to fire over the propeller arc. This was required due to lack of an effective synchronizer.

Right: A Deperdussin TT with crew illustrates the position the gunner had to assume to use the flexible gun. The standing gunner and metal shield the gun was mounted on created a great deal of drag and turbulence, reducing the aircraft's speed and its effectiveness as an interceptor.

monoplane addressed the problem by having the gunner in the front cockpit stand up to fire a flexible gun mounted high enough that it could fire over the propeller arc. Of course, the increased drag of this installation and the turbulence caused by the standing gunner had a significant effect on the aircraft's speed and stability during combat, reducing

Left: The Vickers F.B. 5 "Gun Bus" was a pre-war concept that attempted to create an aircraft to intercept enemy airplanes. Like many similar designs, it was a pusher with the gunner in the front cockpit with a flexible machine gun. Reaching the front in January 1915, it was reasonably successful against the unarmed reconnaissance aircraft of the time – if it could catch them. This one was forced down behind German lines.

Above: The British FE.2 served at the front longer than any other WWI aircraft type, making it by far the most successful 'battleplane' design, or armed combat aircraft. After it was outdated in daylight it was used as a night bomber. Above is the later FE.2d with 250 hp Rolls Royce V-12 engine giving it a reasonable performance and the power to carry three guns, a fixed gun for the pilot and two flexible guns for the observer.

Below & Below Right: The FE.2 was used for bombing, reconnaissance, and to escort other FE.2s. Here an FE.2b gunner demonstrates some of the contortions required to defend the aircraft from attack from behind by German fighters. The gunner had no parachute or safety strap to keep him from falling out of the aircraft during combat maneuvers!

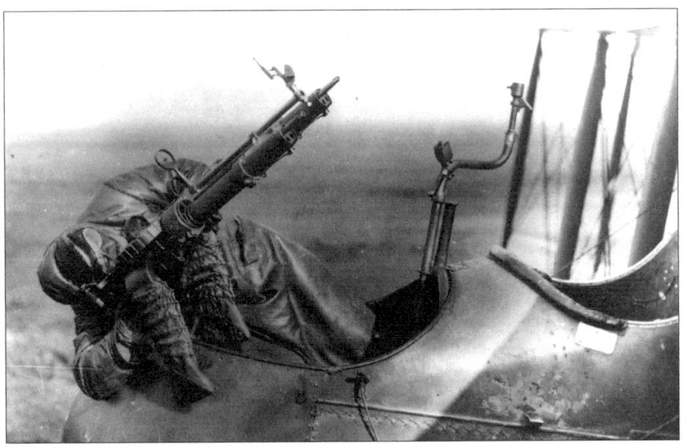

its effectiveness.

Although most Allied 'battleplanes' had one engine, the Germans initially designed twin-engine aircraft for this purpose, again with bombing as one of the expected tasks. The German twin-engine battleplanes initially had a crew of two, with the gunner handling a flexible machine gun in the front cockpit. This was quickly increased to two gunners, the second gunner being situated aft of the pilot to greatly expand the aircraft's field of fire. The twin-engine battleplanes were even less effective than their single-engine competitors because they were slower and less agile. Another problem for Germany

Above: Typical of early Allied reconnaissance planes, this armed Farman MF.11 prepares for take-off from a Belgian airfield in 1915. The crewmen were *Sous-Lt*. Louis Robin & *Lt*. Fernand Jacquet.

Right: The Breguet AG 4 of 1914 was a two-seat fighter; the observer had a flexible gun but there was no gun for the pilot. Despised by its crew when tried in combat, fortunately only two were built.

was inadequate engine production that limited the quantity of twin-engine aircraft they could build.

Germany's twin-engine battleplanes soon found their true combat role as bombers. The British F.E.2 series served as reconnaissance planes, escorts, and light bombers. Successful despite is obsolete pusher configuration, once the F.E.2 became too vulnerable to German fighters during daylight it served until the Armistice as a night bomber. Daylight tactics included sending the F.E.2s out in large groups with most of the aircraft serving as escorts to the aircraft tasked with reconnaissance; if attacked the formation would form a circle, with each aircraft protecting the one in front of it from attack from behind by German fighters. In later years this tactic became known as a 'Lufberry Circle.'

Above: The Dufaux C1 was a radical French design that placed the engine and propeller in the center fuselage to allow a clear field of fire forward for the gunner, who sat alongside the pilot. Amazingly, it was tested in combat.

This Page: The extraordinary British F.E.3 2-seat pusher was powered by a 100 hp Chenu engine. To reduce drag it was designed with a single tail boom that passed through the hollow shaft of the propeller. Built in 1913, it was another pre-war attempt to create a fighting plane without benefit of synchronizer. It was used as a static test-bed for experiments firing a one-pounder gun from the Coventry Ordnance Works but did not go into production and, perhaps fortunately for airmen, did not see war service.

Above & Below: The Voisin 4 was too slow and cumbersome to be an effective interceptor, and the single-shot 37mm cannon was not suitable for air-to-air use; it was more effective for ground attack.

Above: The bizarre B.E.9, which was developed from the B.E.2, fortunately remained a single prototype. See also pages 41 & 276.

Below: The SE.10 proposal was based on the SE.5 airframe but fortunately was not built.

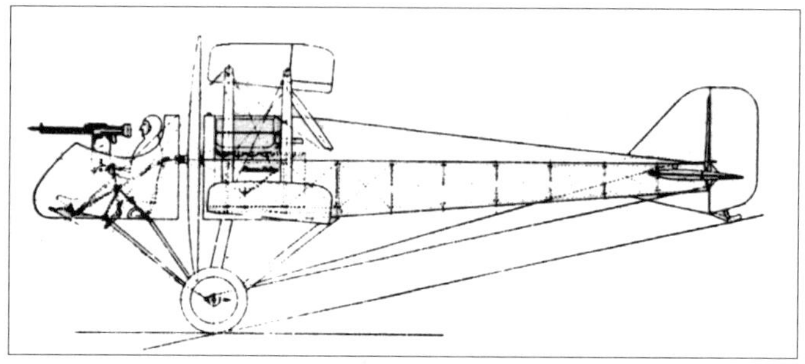

Below: Blériot Parasol flown on scouting missions by *Capitaine* De Marancourt in 1915. The parasol configuration was intended to improve the crew's downward field of view.

In early 1914 key German aviation authorities, including *Idflieg*, the *VPK* (*Verkehrstechnische Prüfungs Kommission* = transport technical investigation commission) and aviation industry executives, discussed the role of military aircraft and reached a consensus. The *VPK* then issued a directive outlining the tactical role of aircraft and specified three categories: *Typ I* was a fast two-seater intended for extended flights for reconnaissance and light bombing; *Typ II* was a light, maneuverable two-seater for short flights over the lines and armed for self-defense; and *Typ III* was three-seater designed to carry a large payload and fly low within range of enemy fire. The *Typ III* was required to have a speed over 120 km/h, climb to 800m in 10 minutes, have a flight duration of 6 hours, and a useful load of 450 kg. In essence the *Typ III* was a battleplane (*Kampfflugzeug* in German).

The German army approved the *VPK* recommendations on 28 April 1914, and demanded that these airplanes be developed as soon as possible. The *Typ I* was essentially the B-type two-seat biplane, of which numerous designs were available, while the *Typ II* eventually converged with the developed B-type to become the C-type armed two-seater. In parallel, a number of companies responded with designs to the *Typ III* requirements.

Several German manufacturers, including AEG and Gotha, designed twin-engine aircraft to the *Kampfflugzeug* concept. The Roland G.I and LVG G.I were built as single prototypes, but the AEG K.I served as the basis for the later AEG G-types that soon were used primarily as bombers. The Gotha G.I was also placed in limited production (18 in addition to the prototype), but like other twin-engine aircraft soon proved a failure in the battle plane role and was increasingly used as a bomber. Later Gotha designs owed little to the distinctive Gotha G.I.

Above: Rudolph Berthold, later to become a successful fighter pilot who scored 44 victories and was awarded the *Pour le Mérite* (aka 'Blue Max'), sits in the cockpit of AEG G.II G.26/15 that he flew while serving with *FFA* 23. Three gunners, each armed with a flexible machine gun, made this G.II a true, operational *Kampfflugzeug*.

Left: The AEG K.I designed to the flawed *Kampfflugzeug* concept was the first twin-engine AEG design. All subsequent AEG G-types followed the same basic configuration. The K.I was a compact design with good handling qualities.

Above: The Rumpler 4A15 was in response to an *Idflieg* request for a *Kampfflugzeuge*. Powered by two 150 hp Benz Bz.III engines mounted as pushers, it set the basic configuration for all subsequent Rumpler bombers. Rumpler G.IIs successfully escorted two-seaters on Eastern Front bombing missions.

Above: The Roland G.I was a prototype *Kampfflugzeug.* A single, fuselage-mounted engine drove two propellers supported by a multitude of drag-producing struts. The gunner in the front cockpit demonstrates his flexible machine gun. The simple landing gear appears to be effective in reducing nose-overs. A slow aircraft with a complex power system, only a single Roland G.I was built.

Left: Designed as a *Kampfflugzeuge*, only one Schütte-Lanz G.I was built. The gunner occupied the front cockpit and had a single flexible machine gun. The fuselage possessed an unusual cross section and heavy supports carried the propeller drive shafts aft so the propellers cleared the wing trailing edge.

Above: The LVG G.I was designed as a *Kampfflugzeug*, and a single example was built in 1915. The arrangement enabled the gunner to fire in a 360° arc above the aircraft and gave a wide field of fire forward, but the standing gunner undoubtedly created a great deal of drag, limiting the aircraft's speed. The *Kampfflugzeugs* (battle planes) of all types soon proved failures as interceptors and air superiority aircraft because they were simply too slow and cumbersome to intercept and engage enemy aircraft. Operational experience soon demonstrated that bombing was the true role of these aircraft, and the early operational *Kampfflugzeugs* were the ancestors of German twin-engine bombers of WWI.

Above: The battleplane heritage of the Gotha G.I is demonstrated here. Designer Oskar Ursinus handles a 20mm Becker cannon in the nose turret while the second gunner aims a Parabellum machine gun; the pilot is seated aft.

Gotha G.I 42/15 *FEDORA*, from the second of three G.I production batches, was delivered to *FEA* (*Flieger Ersatz Abteilung* = Aviation Replacement Unit) 3 in Autumn 1915 before being sent to the Eastern Front.

Right: A Gotha G.I aircraft on the Gotha airfield ready for flight testing. The Gotha G.I epitomized the *Kampfflugzeug* concept, but combat experience soon revealed the flaws in the theory and bombing began to be emphasized instead of air combat. Gradually the G-type aircraft assumed their true role as bombers.

Facing Page: Designer Oskar Ursinus demonstrates the use of the forward gun in a Gotha G.I. The streamlined bomb container between the wheels is clearly visible and the nose of a 20-kg Carbonit bomb just protrudes from the lower front. Together with the bomb-dropping chute beneath the gunner's cockpit the bomb container is a clear indication of the evolution of the battle plane into a bomber. The propellers rotate in opposite directions to minimize torque effect.

Above: This macabre-marked Farman MF.11bis flown by *Capt*. Fernand Jacquet and *Lt*. Louis Robin, *1ere Esc. de Chasse*, was one of several pushers Jacquet flew to become the first Belgian ace.

Above: The unarmed, two-seat Morane-Saulnier L parasol was preferred for scouting due to its good downward vision. It was also built under license by Pfalz. Reginald Warneford also used one to destroy Zeppelin *L.Z.37* by air-to-air bombing, the first Zeppelin to be destroyed in the air by an airplane, for which he was awarded the Victoria Cross.

Above: The unarmed, single-seat Morane-Saulnier H was used for visual reconnaissance (scouting). Also built under license by Pfalz, it became the basis for the Pfalz *Eindeckers*.

Above: This Voisin 3 two-seat pusher carried Lance bombs and a flexible machine gun operated by the observer in the rear cockpit. The machine gun was more effective against aircraft than the single-shot cannon in the Voisin 4.

Below: The unarmed, two-seat Fokker A.II (Fokker factory type M5L) was the immediate predecessor of the Fokker *Eindecker* fighters. Used for visual reconnaissance (scouting), this A.II was crewed by Graf von Holck.

Above & Below: The Mann & Grimmer M.1 in original form (above) and final form (below) with vee undercarriage, propeller shafts braced by struts, and larger rudder. Construction began in September 1914 and first flight was made on 19 February 1915. The 100 hp Anzani radial was mounted backwards at the front of the fuselage and drove two wing-mounted propellers via a long drive shaft, gearbox, and chain drives. This unusual configuration was chosen to give the forward crewman a clear field of fire despite lack of a synchronizing gear. Poor performance caused the designers to replace the original 100 hp engine with a 125 hp Anzani engine, but performance was still poor and the aircraft was eventually wrecked due to a forced landing caused by gearbox failure. Further development was abandoned.

Ironically, the Edward brothers patented a synchronizing gear in 1914 (Patent No. 23790 A.D. 1914) and sent a working model to the British War Office. However, there was no response and the synchronizer was lost to memory, with the result that aircraft like the Mann & Grimmer M.2 and deHavilland DH.2 had to be built without the benefit of this critical device.

This Page: A 1913 design, the French Dorand Armored Interceptor had two 80hp engines coupled together to power both propellers in case of an engine failure. There was a pilot in the middle with gunners fore and aft. Development of armored aircraft was abandoned because of poor performance due to the armor's weight.

Vickers FB.5 "Gunbus" #1623 of the RFC in early markings.

The First Fighters

Before WWI started, machine guns were generally viewed as the weapon of choice for air-to-air combat; the next question was how to mount them. Interestingly, Franz Schneider, a Swiss engineer working with Germany's LVG company, patented a synchronizing device in 1912. France's Robert Esnault-Pelterie also patented a machine-gun synchronizing device before the war, as did the XXXX brothers in England. Surprisingly, given the interest in armingaircraft, none of these inventions was taken up before hostilities!

Initial wartime sorties focused on reconnaissance, the key aviation mission throughout the conflict. A few flyers dropped small bombs or grenades on enemy troops by hand, but this was harassment, not a serious military threat. At first there was essentially no air-to-air combat. Opposing flyers sometimes waved at one another if they passed during a reconnaissance mission, but gradually this collegial attitude changed and hotter heads prevailed. Crews began to exchange pistol or rifle fire when they met. This was almost always ineffective and served only to vent hostility.

Garros and the Armored Propeller

While the aviation establishment struggled to come to grips with air combat, a renowned pre-war French pilot, Roland Garros, took the initiative. Famous for being the first person to fly across the Mediterranean in 1913, Garros had volunteered to fly and was assigned to *Escadrille* M.S. 23. In December 1914 Garros was in Paris and visited Raymond Saulnier, co-owner of Morane-Saulnier, manufacturer of the airplanes with which Garros's unit was equipped. Saulnier had been working with early synchronizers and found that even if the synchronizer was working, sometimes the gun fired late because defects and variations in the bullets made some fire later than others. Sometimes the result was a bullet-hole through the propeller, which could cause the propeller blade to break off, creating extreme vibration that could easily tear the engine out of the airplane, a catastrophic event. Saulnier had experimented with fitting metal bullet deflectors on the propeller to protect it from this disaster.

Meanwhile, Garros had become exasperated with the difficulties of trying to maneuver his airplane into position so his observer could get a good shot at a German aircraft with a rifle and requested permission to test Saulnier's invention in flight. Many French authorities had strong misgivings about the practicality of precisely maneuvering an airplane in flight so as to aim the entire airplane at an enemy airplane, but Garros soon received permission. When Garros and his mechanic, Jules Hue, examined Saulnier's deflectors in more detail, Hue determined that the deflectors were too weak. Hue designed a stronger device attached to the propeller and its shaft, and fitted the modified propeller and a Hotchkiss machine gun to a standard Morane-Saulnier L parasol monoplane.

Poor flying weather delayed a combat test until 1 April 1915. Spotting an unsuspecting German two-seat reconnaissance plane, Garros swung in behind it. The German crew was shocked as Garros opened fire with the Hotchkiss, and after a chase, with the German airplane desperately trying to escape, it fell in flames. Two weeks later Garros scored again, and on 18 April a third fell; the fixed machine gun had proved itself. Later that day, while attacking a train behind the lines, Garros was downed by ground fire that severed his airplane's fuel line. Garros tried to burn the airplane and escape, but he was soon captured and the partially burned airplane, complete with its modified propeller and machine gun, was shipped back to Germany for technical analysis.

Fokker and the First True Fighter Planes

When the Germans captured Garros's aircraft they discovered the steel deflecting plates and immediately grasped their significance. The German army quickly distributed photographs and supporting information to the aviation industry with the intention of replicating Garros's crude device. The story now becomes unclear and, with the destruction of German records during WW2 bombing raids, the full story is unlikely to ever be known. What is clear is that Fokker quickly stepped forward with a working synchronizing gear. In his autobiography *Flying Dutchman* Fokker claimed to have created a working prototype within 48 hours of being shown details of Garros's device. This claim is certainly exaggerated, but Fokker and his team did produce a working device, which they may have been working on for some time before Garros's misfortune. The only existing documentation about the timing of this development was dated May 16, 1915, when the Bavarian liaison officer to *Idflieg* stated in a report that a firing trial was scheduled for May 19 or 20.

Ironically, Fokker adapted the new synchronizer to his M 5K monoplane that was similar in

Above: Morane-Saulnier N with un-synchronized machine gun mounted; the metal deflector plate on the propeller blade is clearly shown.

Below: The metal deflector plate mounted on a Morane-Saulnier propeller blade.

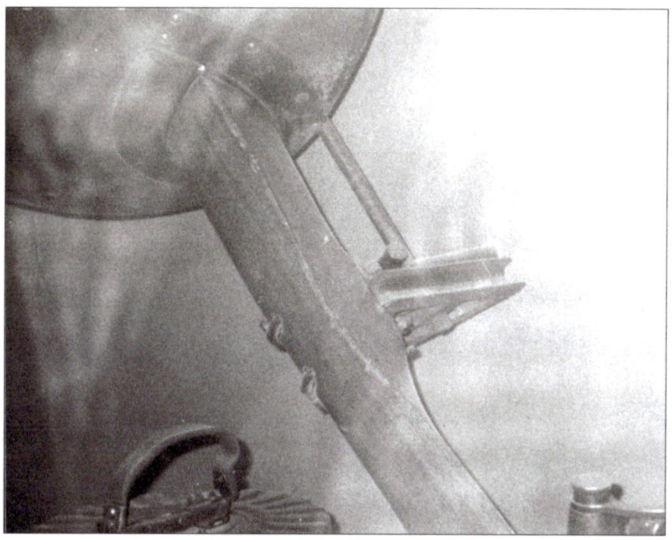

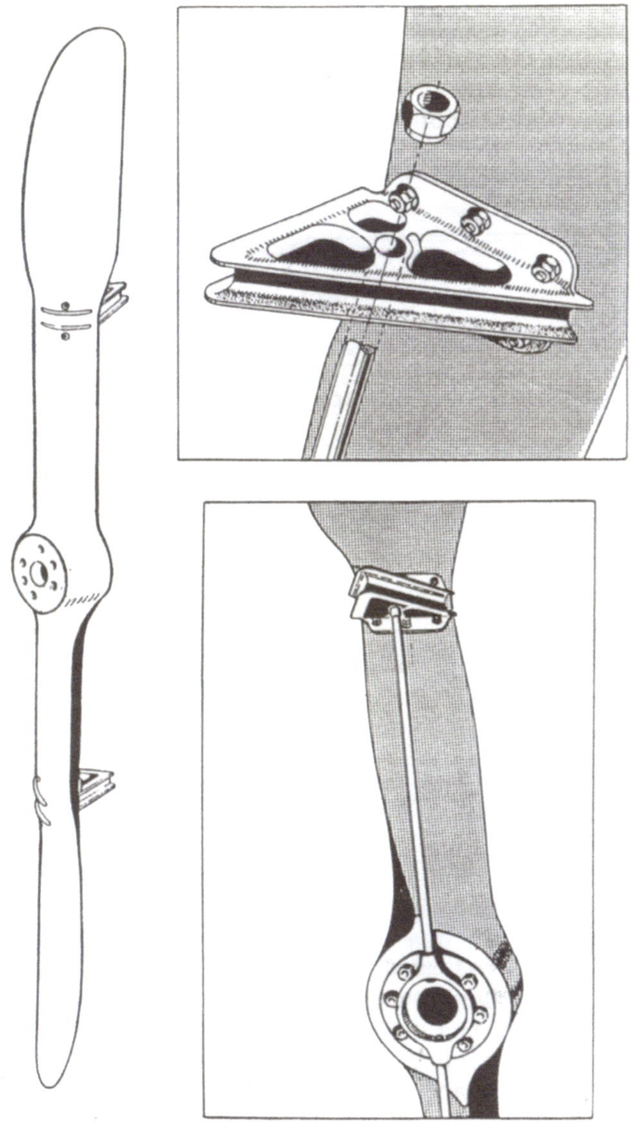

Above: Drawing of the metal bullet-deflector plates as mounted on a Morane-Saulnier propeller.

configuration to the Morane-Saulnier H; the modified airplane was given the military serial E.1/15. This aircraft arguably was the world's first true fighter aircraft, an aircraft designed as a fighter.

Soon *Ltn.* Kurt Wintgens shot down a Morane-Saulnier over French lines to prove the concept worked. A milestone had been achieved and the nature of air warfare changed forever.

The First Fighter Pilots

Two of the Fokker *Eindeckers* (monoplanes) as they were known were assigned to *Ltns.* Max Immelmann and Oswald Boelcke, and on 1 August they attacked a formation of British airplanes that had just bombed their airfield at Douai. A severe gun jam forced Boelcke to land to repair it, but Immelmann attacked. After several jams that he was able to clear, Immelmann succeeded in downing the British airplane. This combat, four months to the day after Garros's first victory, started the so called 'Fokker Scourge,' a time when the Fokker fighters maintained air superiority over Allied airmen, many of whom came to feel like "Fokker fodder".

Immelmann and Boelcke are two names that live on in history. Boelcke became known as the father of German fighter aviation and mentor of the Red Baron, and Immelmann became known for the aerial maneuver that still bears his name.

If the Allies did not immediately recognize the peril they were in, they soon did. By September the situation in the air became critical, and the commander of the RFC's Third Wing ordered that reconnaissance aircraft be escorted by a Vickers

Above: Joseph Frantz and Louis Quenault by the tail of their Voisin after scoring what is thought to be the world's first air-to-air victory on October 5, 1914. Frantz was a pre-war pilot and Quenault was his mechanic. When the war started Frantz was assigned to V.24 and Quenault followed him and became his observer. Some Voisins of V.24 were fitted with a Hotchkiss machine gun for the observer, who, unusually for a pusher design, sat behind the pilot. On Oct. 5, Frantz was able to catch up with a German two-seater from behind and Quenault fired 47 rounds before the Hotchkiss jammed; he then fired a few more shots with his rifle. The German observer fired back with a carbine, but the pilot was killed and the plane crashed, killing the observer as well.

Above: Roland Garros was a famous pre-war French pilot. He won a number of air races, established a world altitude record of 4,250 meters in 1911, and was the first to fly across the Mediterranean Sea in 1913. Garros was in Germany preparing for a flight competition when war broke out; he escaped to France via train before the German authorities could find and intern him. Serving with M.S.23, Garros performed typical reconnaissance and light bombing missions, but at heart was a fighter pilot and helped develop the metal deflector plates to enable an un-synchronized machine gun to be used on a Morane-Saulnier. Returning to the front with M.S.26, Garros downed three German reconnaissance airplanes while still performing bombing and reconnaissance missions. Brought down by ground fire, he was made POW, but finally escaped in February 1918. After returning to France and re-training, he returned to his old unit, now SPA.26 in recognition of its equipment with Spad 13 fighters. The air war had changed dramatically during his captivity, and on October 5, 1918, he was shot down and killed in action.

Left: Although the victory by Frantz and Quenault is generally considered to be the world's first air-to-air victory, it must be noted that on August 26, 1914 Russian pilot Pyotr Nesterov, the first pilot to loop an aircraft, rammed a German Albatros two-seater over the Eastern Front; all three men were killed in the resulting crash.

Above: Jules Vedrines in the cockpit of his Morane-Saulnier N before a sortie; the metal deflector wedge on the propeller is clearly seen together with the un-synchronized Hotchkiss machine gun.

Above: Jean Navarre in the cockpit of his Morane-Saulnier N. The aircraft was fitted with an un-synchronized Hotchkiss machine gun with metal deflector wedges on the propeller. The M-S N debuted in June 1914, just before the war.

The Synchronizer

Ironically, patents for synchronizing mechanisms had been granted in France, Germany, and the UK before the war, but not of these critical devices were taken up at the time! Partly this was due to the prevailing notion of nautical combat between aerial cruisers. Fokker was the first to develop a synchronizer and install it on an aircraft, his M5L, creating the world's first factory-built fighter plane as we know it today. These images from the Fokker factory depict some of the steps in Fokker's synchronizer development.

Above: Test-bed showing the critical components of the Fokker synchronizing gear.

Above: Test-bed for the Fokker synchronizing gear. This may have been the setup demonstrated before German officials that resulted in the *Idflieg* order for the first Fokker fighters.

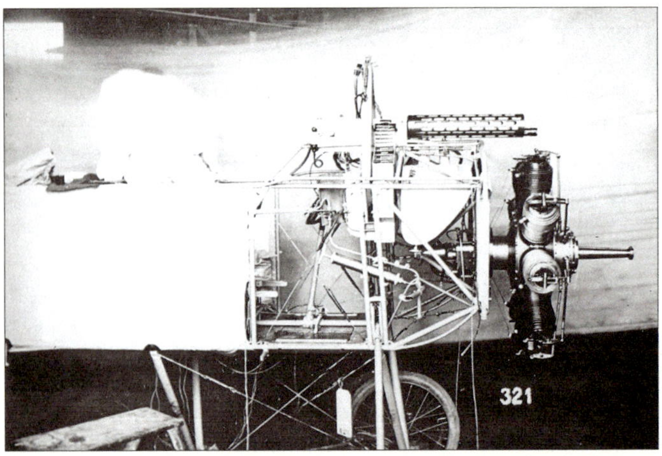

Above: A fixed machine gun synchronized by the Fokker mechanism installed in a Fokker E.I fighter.

Below & Right: Disk and propeller test rig.

Right & Below: Two Sanke card photos of Anthony Fokker with his M5L design. The unarmed M5L (L for *Lange*, or long wing) was the basis for the A.II unarmed two-seat reconnaissance aircraft and the M5K (K for *Kurz*, or short wing) sired the armed A.III single-seat fighter, soon re-designated E.I.

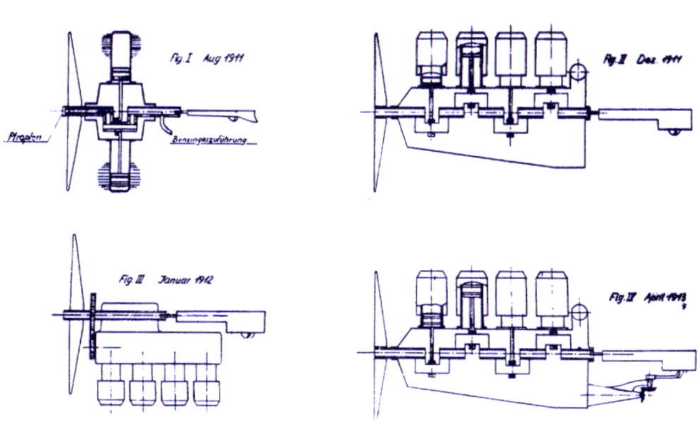

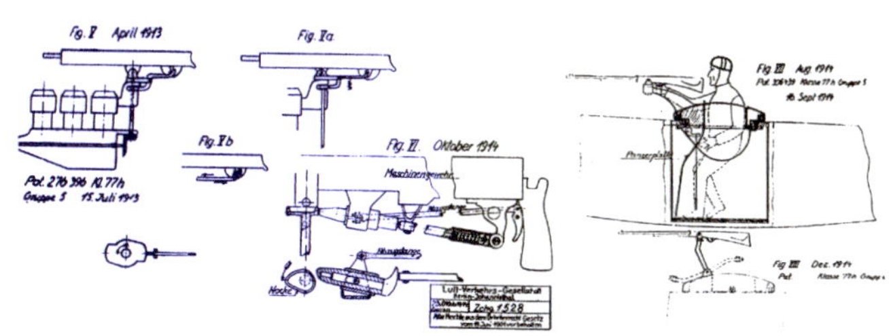

Left & Above: Pre-war gun synchronizer and gun turret drawings by Franz Schneider. The sketch at the top left was made in August 1911, and designs continued until the drawing at top right in August 1914. Aircraft armament was an important pre-war topic and synchronizers were patented in France, Germany, and Great Britain before the war, but none were developed or deployed.

Above: Fokker E.I at the Fokker factory. The E.I was powered by a 7-cylinder, 80 hp Oberursel U.0 rotary engine.

Below: *Ltn.* Kurt Wintgens, first pilot to score a victory using a Fokker E-type fighter and an early ace, sits in his Fokker E.I 5/15. The black rudder and black and white fuselage bands were typical of *FFA* 48 where Wintgens had been serving before returning to *FFA* 6b. A headrest was provided to help steady the pilot's head when he was aiming the gun.

Gunbus. Although the Fokkers were faster and more maneuverable than the Gunbus, the Vickers, with its flexible gun, was surprisingly successful in combat with them, and from then on German pilots tended to describe all Allied pusher fighters as 'Vickers'.

The Fokker itself was not a wonder airplane; for the time it had good speed, climb, and maneuverability, but its main attribute as a fighter was its synchronized machine gun. Indeed, both the Morane-Saulnier N and the Bristol Scout had better performance than the Fokker, but neither had a synchronized gun. The Morane relied on deflector plates as had Garros, and the Bristol Scout had a variety of gun mounts that aimed the gun to miss the propeller. The non-synchronized gun mount that found most favor, and continued in use to the end of the war, was mounting the gun above the upper wing to fire over the propeller arc. The drawback was the gun was out of convenient reach if a jam occurred, which gave pilots many awkward moments during air combat.

The Allies did not deploy a fighter with synchronized gun until the Summer of 1916, when several types appeared. Until then, the Allies had to

Above: Max Immelmann in Fokker E.I 13/15. Oswald Boelcke used this aircraft to give several visitors rides in early August 1915!

Right: Ernst von Althaus, another early ace to win the *Pour le Mérite*, sitting in a Fokker E.I. The E.II and E.III used essentially the same airframe as the E.I, making it difficult to distinguish between the types. The key difference was the E.II and E.III used the more powerful 9-cyliner, 100 hp Oberursel U.I rotary. The E.III had more fuel and oil capacity than the E.II but the differences were not externally visible; some E.IIs were converted to E.IIIs.

Below & Right: Two views of Fokker A.16/15, one of the earliest armed *Eindeckers*, assigned to *Lt*. Parschau. The early gun with stock, the external chute for ammunition, and the headrest to steady the pilot's head are all visible.

Above: An iconic image of a Fokker *Eindecker* in flight.

rely on pusher fighters and those, like the Nieuport, with an over-wing gun.

Development of Fighter Tactics

The Fokker pilots developed tactics that, in elaborated form, are still used today. A Fokker pilot would cruise over the front and, upon spotting an Allied airplane, would attempt to position himself above it and between the sun and the target, which would make him invisible to the crew. Waiting for the right moment, the Fokker pilot would then dive down from behind the enemy airplane to fire a devastating burst from close range. If the enemy plane did not fall immediately, the Fokker pilot would continue his dive, zoom back to altitude, and repeat the attack. If the enemy was an armed two-seater, the Fokker would initially dive beneath it to avoid the observer's defensive fire, using the fuselage of the enemy plane to block visibility and return fire, then zoom up underneath it to fire a burst from close range into its belly. These were two variations of the 'Fokker bounce'. If the Fokker did not have an altitude advantage, the pilot might stalk the enemy airplane from behind and below so its crew would not see him approach. Of course, the Allies adopted the same tactics when they fielded fighters with fixed, forward-firing guns.

Importantly, the Germans were allocating the Fokkers in ones and twos to reconnaissance units instead of grouping them into squadrons. Although they continued to win victories and had air superiority, this failure limited their effectiveness. Certainly the limited number of Fokkers also played a part, but production could have been increased. The fact that it was not, together with failure to group the Fokkers into squadrons, gave the Allies the opportunity to continue their reconnaissance behind German lines despite increasing losses. The RFC responded by arming all reconnaissance planes with defensive machine guns and escorting them with one or two airplanes. But Trenchard, the commander of the RFC in France, avoided large changes in tactics that he feared might make the Germans respond with better tactics of their own, such as grouping their fighters into squadrons. Trenchard realized the RFC would incur more losses, but he wanted to continue offensive operations while waiting for improved British fighters.

Nevertheless, by its peak in January 1916, the Fokker Scourge had severely limited RFC operations by forcing the RFC to escort each reconnaissance airplane with at least three other armed airplanes. It thus took many more airplanes to fly each reconnaissance mission, greatly stretching RFC resources and limiting the number of reconnaissance missions that could be flown.

Above: Morane-Saulnier I with synchronized gun.

First Generation Fighters	At Front
Bristol Scout	9/14–9/16
de Havilland DH.1	1916–1917
FE.2a	5/15–1/16
FE.2b/2d[1]	11/15–8/17
Fokker *Eindeckers* (E.I, E.II, E.III, E.IV)	6/15–12/16
Morane-Saulnier L	3/15–10/15
Morane-Saulnier I, N, V	6/15–10/16
Martinsyde S.1	11/14–6/15
Pfalz *Eindeckers* (E.I, E.II, E.III, E.IV, E.V)	10/15–12/16
Siemens-Schuckert E.I	10/16–4/17
Sikorsky S.16	3/16–9/17
Sopwith Tabloid	8/14–4/15
Spad SA.1, SA.2, & SA.4	9/15–5/17
Vickers F.B.5/F.B.9 Gunbus	2/15–6/16

Note 1: Once it was removed from day combat due to its increasing vulnerability, the FE.2b/2d served as a night bomber until the Armistice.

The first generation fighters and the dates they were at the front are listed in the table above. These airplanes either used airframes designed before the war or, in the case of the Vickers, were built to a pre-war concept. The Vickers is included in the list because it was operationally important, but it was built to the two-seat battle-plane concept and in that regard was not a true fighter plane, hence the asterisk. The other aircraft are all single-seaters.

The *Frontbestand* table below shows the front-line inventory of the various German monoplane fighters, known as E-types, in detail. These aircraft all featured synchronized machine guns; most carried only one gun but the Fokker E.IV and Pfalz E.IV were both powered by a 160 hp two-row rotary and carried two guns.

Before the war Pfalz had licensed two Morane-Saulnier designs. The Pflalz E-types were developed from their licence-built versions of the Morane-Saulnier H. This pre-war airframe was not built to withstand the stresses of maneuvering combat and eventually the Pfalz E-types were removed from the front due to structural failures.

Of the Allied types listed above, only the Morane-Saulnier I and V, more powerful developments of the M-S N, had synchronized guns. Only four M-S I fighters were built for the RFC and no more than 20 for Russia; only a dozen M-S V fighters were built for the RFC. The Morane-Saulnier I and V were slightly faster than the Fokker E.I–E.III, but were few in number and the basic design had challenging handling qualities.

Despite their limitations, the Fokker monoplanes were the best fighters of the first generation; they had the best armament, decent performance, and were reasonably maneuverable and easy to fly. The Bristol Scout had a somewhat better airframe but was severely limited by its make-shift armament.

Frontbestand Inventory of German Monoplane Fighters at the Front

		1914			1915						1916						1917						1918			
Manufacturer and Type		31 Aug	31 Oct	31 Dec	28 Feb	30 Apr	30 Jun	31 Aug	31 Oct	31 Dec	28 Feb	30 Apr	30 Jun	31 Aug	31 Oct	31 Dec	28 Feb	30 Apr	30 Jun	31 Aug	31 Oct	31 Dec	28 Feb	30 Apr	30 Jun	31 Aug
Fokker	E.I						4	17	23	26	12	22	3	1												
	E.II							2	8	14	20	13	4	5	2	1					2					
	E.III							3	23	40	67	110	101	64	28	11	2	1	1		8	7				
	E.IV								1	6	6	28	29	25	16	2										
	E.V (D.VIII)																									80
Pfalz	E.I								2	14	26	27	13	11	3	1										
	E.II									6	11	20	30	17	3											
	E.III								1	1		4	8													
	E.IV											5		1												
	E.V												3													
SSW	E.I														5	2	1	1								
Unknown				2																						
Total:				2			4	22	58	107	142	229	191	124	57	17	3	2	1		10	7				80

Above: The *Frontbestand* inventory of German E-type fighters at the front shows the dominance of the Fokker fighters, which arrived at the front first and served in greater numbers than any other *Eindeckers*. Pfalz was second to the front yet disappeared from the front earlier than the Fokkers due to structural weakness that caused their removal for safety reasons. The most likely cause of the Pfalz structural weakness was that the Pfalz designs were based on the pre-war Morane-Saulnier H airframe that was not designed to withstand maneuvering combat. The SSW E.I entered service after *Eindeckers* had been surpassed by the DH.2 and Nieuport biplane fighters; most served as trainers. The two unknown E-types listed at the end of 1914 are an error.

Above: Oswald Boelcke after being awarded the *Pour le Mérite* for scoring eight victories.

Above: Max Immelmann after being awarded the *Pour le Mérite* for scoring eight victories.

On the morning of January 12, 1916 *Hptm.* Oswald Boelcke and *Oblt.* Max Immelmann both downed their eighth enemy aircraft. For this unprecedented feat both were awarded the *Pour le Mérite*, Germany's highest award. Only 71 awards of the *Pour le Mérite* had been made in the entire war to that date, and only three to junior officers; the rest had gone to generals, admirals, and royalty. Boelcke and Immelmann, both fighter aces, were the first airmen to win the award and only the second and third *Leutnants* in the entire German military. Both men instantly became national heroes. Not only were their exploits followed from then on by the German media, their achievements were even acknowledged in the magazines and newspapers of their enemies.

This instant acclaim, emulated in other countries about their fighter aces, created the legend of the fighter pilot, and especially the fighter ace, as hero.

The exploits of the fighter aces were widely reported and acted to some degree as an antidote to the horrific news of the endless mass slaughter of the trenches. Few of the countless acts of courage in the trenches could be reported amidst the mass carnage created by artillery and machine guns. On the ground the terrible toll continued without respite or apparent result and despite courage, skill, or sacrifice.

In contrast, the fighter pilot fought in the pure air, not the endless mud of the Western Front, and lived or died by his own skill in knightly combat. Moreover, the achievements of the fighter pilot were easy to quantify; how many enemy airplanes had he shot down?

The difficulty of measuring the accomplishments of reconnaissance and bomber pilots greatly impeded their elevation to honorary knighthood despite the vital nature of the tasks they performed and the similar courage required. This is exemplified in the fact that 62 of the 81 aviation recipients of the *Pour le Mérite* were fighter aces even though only about one quarter of all German pilots were fighter pilots.

Fokker E.II 69/15 of *Lt*. Kurt *Freiherr* von Crailsheim, *FFA* 53

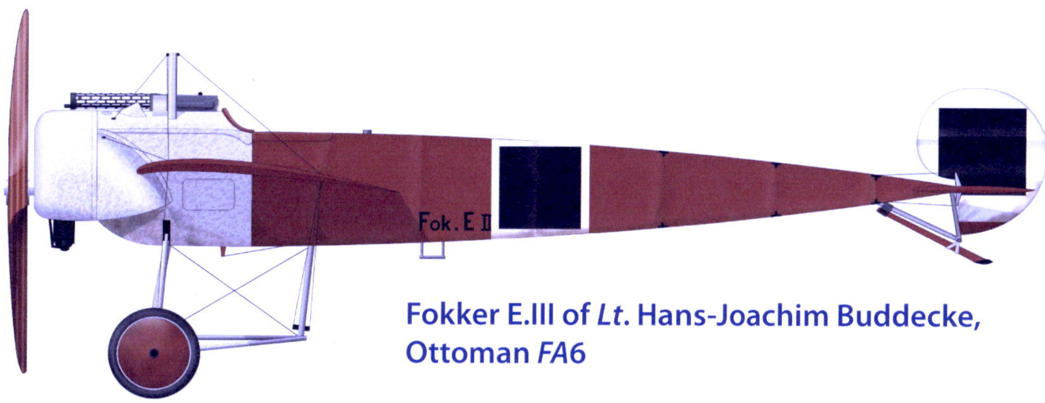

Fokker E.III of *Lt*. Hans-Joachim Buddecke, Ottoman *FA*6

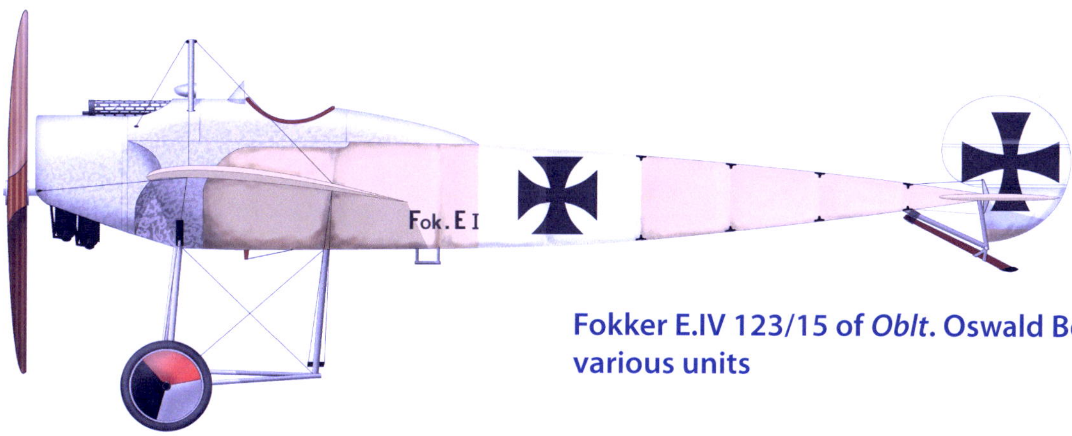

Fokker E.IV 123/15 of *Oblt*. Oswald Boelcke, various units

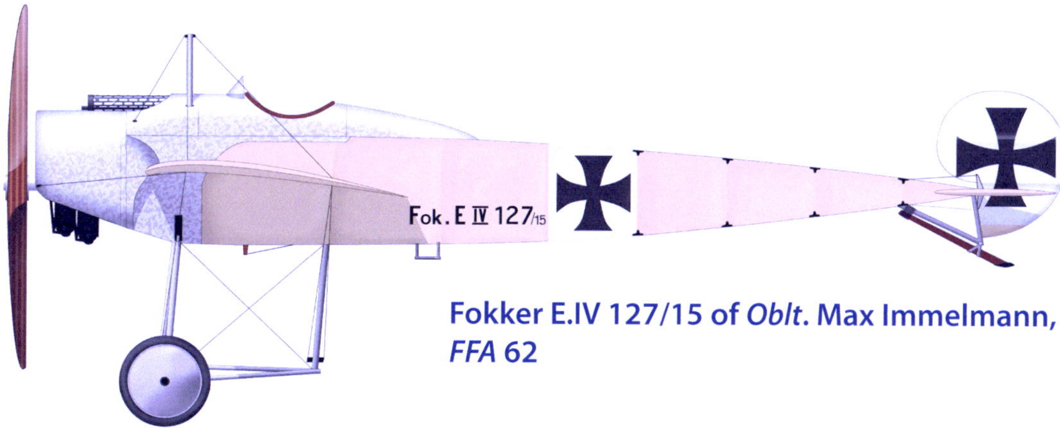

Fokker E.IV 127/15 of *Oblt*. Max Immelmann, *FFA* 62

Above: Hans-Joachim Buddecke in his Fokker E.II 36/15 while *FFA* 23 mechanics get ready to start the engine.

Above: Hans-Joachim Buddecke and a mechanic in front of a Fokker E.III in Turkish markings in the Middle East.

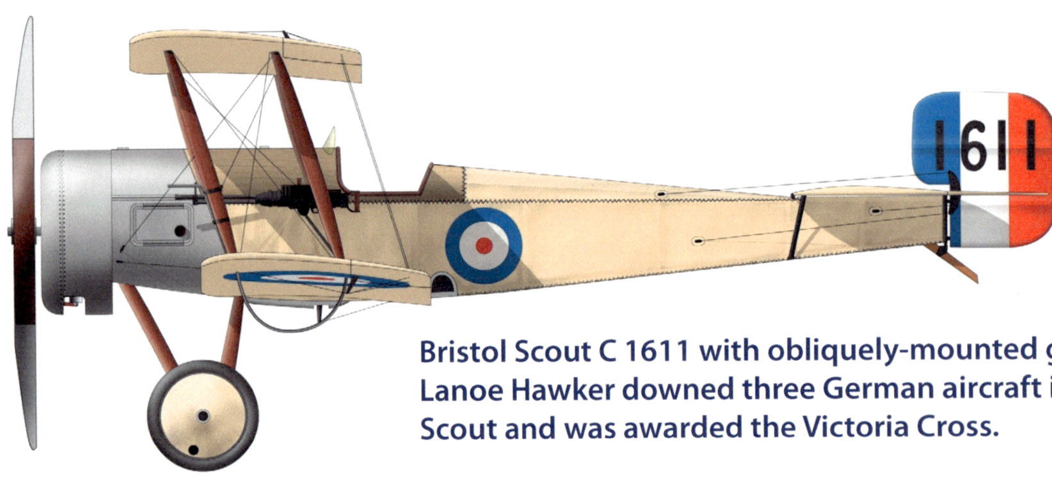

Bristol Scout C 1611 with obliquely-mounted gun. Lanoe Hawker downed three German aircraft in this Scout and was awarded the Victoria Cross.

This Page: The Bristol Scout was designed as an un-armed, high-speed single-seater for scouting. Later some aircraft were fitted with armament. Lack of a synchronizer meant Scout 1611 (below) had to avoid firing through the propeller arc, and an oblique-firing gun was fitted. Major Lanoe Hawker (above) downed three German aircraft with this aircraft to become the first fighter pilot awarded the Victoria Cross, Britain's highest award for valor in combat. Hawker's tactics were simple; "Attack Everything." Above right is a close-up of the gun mounting.

Above: A production E.IV with its two guns mounted on either side of centerline. The prototype carried three guns, but production aircraft only had two. Early aircraft had their guns offset to starboard.

Right: Production E.IV 189/16 with its two guns mounted either side of center.

Below: This E.IV with black and white 'X' markings may have been assigned to *KEK* 3.

Above: A Fokker E.II (foreground) with Pfalz E.I 215/15 and a Pfalz E.II in the background provides an opportunity to compare the fighters. Allied airmen rarely, if ever, distinguished the Pfalz E-types from the Fokker E-types; all were identified as 'Fokkers'. The rudders were different shapes, but the most noticeable difference between the types is the finish; the Fokker is plain linen and metal, whereas the Pfalz aircraft are white with black trim, black cowlings and black forward fuselages. The Pfalz E-types also have crosses on their elevators, while the Fokkers do not. This is the airfield of *Feldflieger-Abteilung* 19 on the Eastern Front, where many of the Pfalz E-types were assigned.

Above: *Lt.* Otto Kissenberth with a Pfalz E.I at *KEK* (*Kampfeinsitzer Kommando*) Ensisheim. The E.I shows the iron crosses beneath the wings and tailplane. The wing structure shows through the fabric, and the Pfalz logo is on the center of the engine cowling. Kissenberth first became a fighter pilot at *KEK* Ensisheim and went on to score 20 confirmed victories and win the *Pour le Mérite* before crashing a captured Sopwith Camel. His last victory was in the captured Camel. His injuries kept him out of further combat. Kissenberth was one of a number of aces who wore glasses.

Above: The Pfalz E.IV was distinctive due to its twin-row 160 hp Oberursel U.III rotary engine, which required an additional front bearing support, and twin Spandau LMG 08 machine guns. Pfalz was fond of the German national insignia and applied it to upper and lower surfaces of the tailplane in addition to its standard locations. This E.IV even has the iron cross on the wheel covers.

Left: This view of an uncovered Pfalz E.IV shows its twin-row 160 hp Oberursel U.III rotary engine and additional front bearing support in more detail.

Pfalz E.I 215/15 of *FFA* 19

Pfalz E.II '5'

Pfalz E.IV of *Lt.* Friedrich Grünzweig, *Feldflieger Abteilung* 9b

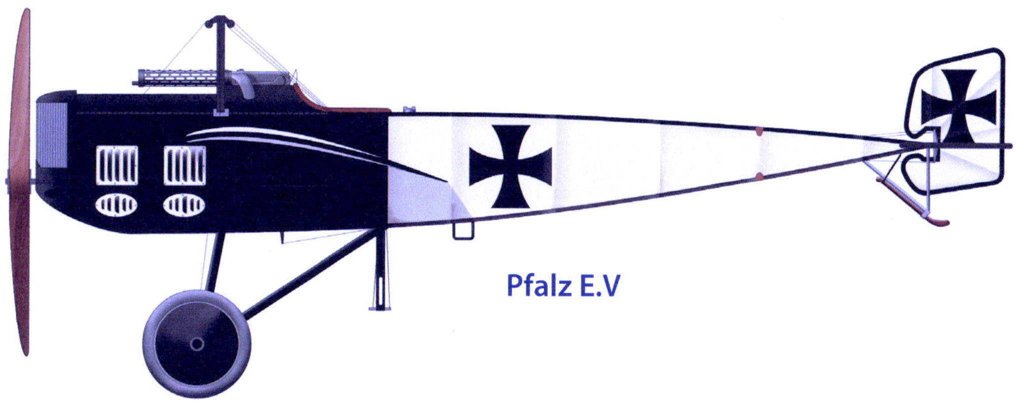

Pfalz E.V

Above: This new Pfalz E.V stands ready for flight on the Speyer airfield. It was the only Pfalz E-type powered by a water-cooled engine, a 100 hp Mercedes D.I. Its streamlining made it faster than the Pfalz E.IV with 160 hp Oberursel U.III rotary, but the basic Morane-Saulnier airframe was incapable of further development. Superior German and Allied biplane fighters were already in service, sealing the fate of the E.V.

Above: Sikorsky S-16ser #204 equipped with a Vickers machine gun and Lavrov's synchronization gear was the first fighter designed in Russia. The S-16 was originally designed as a two-seat trainer and scout with side-by-side seating, but was later adapted as a first-generation fighter. The S-16 could be operated on skis and some had over-wing guns instead of synchronized guns. Power was either an 80 hp Gnôme or 100 hp Monosoupape.

Above: SSW E.I fighters in production at the modern SSW factory; production aircraft dispensed with the small spinner fitted to the prototype. The SSW design was cleaner than the Fokker and Pfalz *Eindeckers* but was simply too late for important operational use at the front and most were delivered to single-seat flying schools.

Above: The sole SSW E.II prototype is shown in side view. The large exhaust for its in-line Argus engine is prominent. The SSW E.II was one of two *Eindeckers* to feature an in-line engine, the other being the Pfalz E.V.

Above: The NFW E.I at Johannisthal. With an 80 hp Oberursel, it was under-powered for a fighter in 1916 when it was built. Its wing design was much more advanced than the Fokker and Pfalz *Eindeckers*, being a two-spar, semi-cantilever design with plywood covering and ailerons instead of wing-warping. The pilot's cockpit was between the wing spars. Only the single prototype was built; the *Eindecker* era was quickly coming to a close when it appeared.

Below: The de Havilland DH.1 powered by a 70 hp Renault appeared in January 1915 and was designed for fighting and reconnaissance. The DH.1A below had a more powerful 120 hp Beardmore. A few were issued to Home Defence units and six were sent to the Middle East in 1916, including the aircraft below. The DH.1A was most significant as predecessor to the smaller and more successful DH.2 single-seat pusher fighter.

Above & Right: The Martinsyde S.1 powered by an 80 hp Gnôme rotary was used in small numbers armed with an over-wing Lewis machine gun.

Above: The Sopwith Tabloid was similar to the Martinsyde S.1 and was powered by the same 80 hp Gnôme rotary. The Tabloid was used in small numbers armed with a Lewis machine gun, some fitted over the wing, others mounted on the fuselage firing at an angle.

Right: Sopwith built a Gunbus too. A 1913 design, this one is armed with a flexible Lewis gun. It is not known if any were flown in combat.

Above & Right: The Spad SA series was another radical approach to an 'aerial cruiser' that served in modest numbers, at least 96 being built. Cooling problems limited the engine power.

Below: Spad SA.2 S.31 scored the only confirmed SA-series victory on the Western Front. The SA.2 and SA.4 were more successful with the IRAS on the Eastern Front, where aerial opposition was not as intense.

Spad SA.2 *Mister Fly* of the French Air Service

Spad SA.4 S.79 of the 30th Corps Detachment, Imperial Russian Air Service

Spad SA.4 of the XIXth Corps Detachment, Imperial Russian Air Service

Above: Developed from the B.E.2, the B.E.9 featured the same 'pulpit' gunner's cockpit configuration as the Spad SA series. Although it was evaluated in combat, it fortunately remained a single prototype.

Above: The original FE.2 was a 1911 re-design of the FE.1 of 1910. The FE.2 was completely re-built in 1913, emerging as a completely new aircraft. This second FE.2 design is shown above and bears only a limited resemblance to the production FE.2 series.

The battleplane or 'aerial cruiser' concept was erroneous and most aircraft designed for that purpose were either unsuccessful or, like the German G-types, evolved into different operational roles. In the end there was one really successful battleplane, the iconic British FE.2 series, which served at the front longer than any other WWI aircraft design: the FE.2a arrived in France in May 1915 and the FE.2b served as a night bomber until the Armistice.

By far the most numerous variant of the FE.2 series was the FE.2b, with observer in front and a 160 hp Beardmore for power. Early aircraft had a single flexible Lewis gun for the observer; later aircraft had two flexible Lewis guns.

The FE.2 compensated for its modest performance with its robust, honest character and the determination of its crews. The aircraft was tough and stable and the crews developed tactics to compensate for its vulnerability to attack from the rear; the aircraft was normally flown in formations during daylight and when attacked the formation would form a circle with each aircraft protecting the tail of the aircraft in front of it.

The FE.2b fought the Fokker *Eindeckers* on basically even terms, but the arrival of the powerful Albatros fighters put the FE.2b at a distinct disadvantage. Nevertheless, the Red Baron himself was shot down and nearly killed in combat with a formation of Rolls-Royce powered FE.2ds.

The FE.2 initially served as a general purpose battleplane, used for offensive patrols, escort, and reconnaissance in daylight until late summer of

Above: Sergeant Thomas Mottershead was the only enlisted pilot to win the Victoria Cross during WWI. On Jan. 7, 1917 his FE.2d of No.20 Squadron was set on fire during air combat. He brought his burning FE.2d home and landed it safely, saving his observer. Unfortunately, Mottershead died of his burns.

Left: Unlike other variants of the FE.2, the FE.2c seated the pilot in front. The intention was to improve the pilot's field of view for night flying, and a few were used in combat by No.25 and No.100 Squadrons.

Above: FE.2b in service as a night bomber.

Right: Flying reproduction of an FE.2b built by Vintage Aviator Ltd. in New Zealand.

1917. The FE.2 also served as a night fighter but with limited success due to its inadequate climb and ceiling. However, the FE.2 was used as a night bomber until the end of the war.

Above: FE.2d A6516 serving with No. 20 Squadron as a day fighter/reconnaissance plane. The pilot has a fixed Lewis in addition to the two for the observer. In this photo the observer is demonstrating how he would use the large, externally-mounted camera in flight. The '2d' had much better performance due to its 250 hp Rolls-Royce V-12.

Above: A few FE.2b aircraft were modified into single-seaters with a one-pounder pom-pom gun for anti-Zeppelin sorties over the UK.

Second Generation Fighters

Above: The de Havilland DH.2 pusher was Britain's key contribution to the second generation of fighter aircraft. The prototype flew in July 1915, just as the Fokker Scourge was getting underway, and the design owed nothing to the subsequent air combat experience – the DH.2 initially featured a flexible gun like a two-seater, and the gun was only mounted in a fixed installation later. Most were powered by a 100 hp Gnôme Monosoupape engine; their Fokker E.II and E.III opponents also had 100 hp rotary engines.

Second generation fighters were intended to carry armament from the outset but were not designed specifically to counter the Fokker *Eindeckers*. They were generally a little more sophisticated in their aerodynamic and structural design than first-generation fighters but at first relied on similar engines. The key was that they were designed as armed warplanes from the outset.

The key second-generation fighter was the French Nieuport 11 sesquiplane, introduced to combat in early 1916. The British DH.2 pusher was also very important in ending the superiority of the Fokker monoplanes, but was limited to British service and had limited growth potential.

Verdun

On 21 February 1916 a bombardment by 850 German guns on the French fortress city of Verdun signaled the beginning of a ten-month battle that became a bloody stalemate. But if the ground battle was indecisive, the evolving aerial battle strongly influenced both sides.

Second Generation Fighters	At Front
deHavilland DH.2	12/15–6/17
F.E.8	8/16–6/17
Fokker D.I, D.II, D.III, D.IV, D.V	8/16–8/17
Mosca MBbis	12/16 – 1917
Nieuport 10, 12, 20	2/15–2/17
Nieuport 11 & 16	1/16–8/17

Unlike the Germans, the French had already started grouping their fighters into units that fought together. Only one fighter *escadrille* was in position near Verdun when the battle started, but within a week six fighter and eight reconnaissance escadrilles were on hand. These were opposed by a mixed force of 168 German aircraft and even four Zeppelins. While not as committed to the offensive in principle as Trenchard and the RFC, the French decided to adopt an offensive strategy of sending groups of three to six fighters to patrol behind German lines, and they asked for, and received, much equipment from the RFC in support. As important, the French fighter

Above: The iconic second generation fighter aircraft was the Nieuport 11, and standing in front of this N.11 is Jean Navarre, its most famous exponent. Navarre was the first French ace to be the 'ace of aces' and for his constant patrolling over the Verdun battlefield became known as "The Sentinel of Verdun". The N.11 sired a long line of Nieuport fighters.

escadrilles now flew the Nieuport 11, a diminutive biplane with an over-wing Lewis machine gun. The Nieuport had a narrow chord lower wing for good downward visibility and was faster and more maneuverable than the Fokker *Eindeckers*. The Fokkers represented the first generation of fighter technology and the Nieuport was a second generation fighter.

In contrast to the French, the Germans adopted an air blockade defensive strategy known as *Luftsperre* that was ineffective and wasteful of resources. This strategy, combined with the new French technical superiority, gave air superiority back to the French, who ruled the skies over Verdun by mid April. Known as the *Groupe des Cigognes* (Stork Group), the French fighter pilots included many stars who became famous aces, such as Jean Navarre, Georges Guynemer, and Charles Nungesser. Due to his constant patrolling in his distinctively painted Nieuport 11, Navarre became known as "The Sentinel of Verdun".

From this time the Germans, realizing they would always be outnumbered by Allied airplanes, stayed on the strategic defensive, and intercepted Allied aircraft over their own lines to conserve their strength. Allied pilots downed behind German lines became PoWs lost to their air service; surviving German pilots returned to their units. The prevailing westerly winds aided the Germans, usually blowing airplanes toward German lines, which could be crucial during air combat.

Above: Jean Navarre patrolling over Verdun in his N.11 with its distinctive blue/white/red fuselage bands.

The air battle over Verdun taught some hard lessons. First, the success of the offensive strategy convinced the Allies that was the most effective employment of airplanes, and they pursued the

Above: The F.E.8 pusher was Britain's second contribution to the second generation of fighter aircraft. It was to the same configuration as the DH.2, was powered by the same 100 hp Gnôme Monosoupape engine, and was armed with the same fixed Lewis machine gun. Not surprisingly, it had similar performance to the DH.2. However, it had the misfortune to reach the front much later, just as the third-generation Albatros fighters were appearing, and it was soon retired.

aerial offensive from then until the Armistice. Second, Germany realized the ineffectiveness of the air blockade and abandoned it. Third, the effectiveness of specialized fighter units operating in strength was clear, and the fighter emerged as a full-fledged partner to the reconnaissance airplane and bomber. Finally, increased cooperation between the British and French air services was maintained from that time onward.

Battle of the Somme

The British had long planned their offensive on the Somme, but by the time they actually launched it on 1 July 1916, much of its intention was to take pressure off the French at Verdun. The first day of the battle was the worst day in the history of the British Army; the British sustained 60,000 casualities for little change in the front lines. The Battle of the Somme ground on until November with massive casualites on both sides but no real change in the ground situation.

However, the Allied situation was much better in the air. The French Nieuport 11 entered British service in January 1916, soon followed by the British DH.2 and FE.2b. Both these types were pushers designed from the outset as fighters. The FE.2b was a pre-war design two-seater with the gunner in front of the pilot with one or two flexible Lewis guns, one firing forward and one back over the upper wing. Fairly maneuverable for a large airplane, the FE.2b was a well-armed and dangerous opponent for the Fokkers. The DH.2 was a single-seat fighter that was superior to the Fokker in every aspect of performance. These fighters quickly established air superiority over the Fokker in the British area of the Front and finally dispelled the 'Fokker Scourage' mentality of the RFC, which had given the Fokker a much fiercer rcputation than it deserved.

While literally holding their ground in the land battle, the Germans were being defeated in the air. German losses rose significantly as the Allies gained air superiority all along the Western Front. Even Immelmann fell in combat with FE.2bs, although it is uncertain whether it was structural failure or British gunfire that caused his death. He had 15 confirmed victories and scored two more the day of his death, for a total of 17. Regardless, with Immelmann's fall, the Fokker *Eindecker* era was over.

Shocked by Immelmann's death, the Kaiser ordered Boelcke, who had 18 victories at the time, away from the front to keep him safely out of combat. By this time the names of Immelmann and Boelcke were famous throughout Germany as heroes of the air. Guynemer and Nungesser were similarly famous in France. In a time of senseless mass slaughter, the home fronts needed heroes, and the fighter aces filled that need. Their individual courage and exploits could be celebrated as a ray of hope and knightly courage and prowess as the war ground on with countless soldiers dying anonymously in the horror of trench warfare.

The bloody Somme stalemate had other

Above: Jean Navarre in a Nieuport 16. Using the N.11 aircraft, the N.16 was fitted with the heavier and more powerful 110 hp Le Rhône instead of the 80 hp Le Rhône used in the N.11. This made the N.16 faster and gave it a better climb rate at the expense of poorer handling qualities due to being somewhat nose heavy.

implications. For one, the last hope for a negotiated peace died with the tens of thousands of casualties. Also, Germany recognized that Britain was a more serious threat than France, and the bulk of the fighting shifted to the British sector. This was particularly true in the air, where Trenchard's relentless offensive strategy resulted in far more air combats than with the French. This is hardly to suggest that air combat with French airplanes was easy or unimportant; it was just that French aviation was not consistently on the offensive like the RFC and RNAS.

Above: The two-seat Nieuport 10 was designed as a reconnaissance plane powered by the 80 hp Le Rhône and was the first Nieuport to use the sesquiplane wing. Later aircraft added a gun for the observer and some had a gun for the pilot as shown here. The N.11 was a smaller, single-seat development of the N.10.

Left: This N.21 had an N.23 fuselage with 80 hp Le Rhône and N.17 wings; this one has American rudder stripes.

Above: The Weymann W.1 pusher fighter of 1915 was an innovative French attempt to create a low-drag pusher fighter without the need for synchronizing gear. Powered by an 80 hp Clerget in the middle of the fuselage driving the tail-mounted propeller via a transmission and extension shaft, it carried two fixed, un-synchronized guns and had a metal airframe. Flight testing is thought to have been abandoned in late 1915.

Left: A heavily-armed N.10 serving with the IRAS (Imperial Russian Air Service). The N.10 was built under license in Russia by the Dux and Lebedev plants. With only 80 hp the N.10 was underpowered for a two-seater, especially when armed.

Left: To overcome the lack of power in the 80 hp N.10, an enlarged, more powerful version with 110 hp Clerget was built as the Nieuport 12. The N.12 was built under license in Britain and this aircraft was one of those as indicated by the supplementary vertical fin.

Right: This well-known photo shows Cdr. Salmson, RNAS standing in front of a N.10 with gun fixed to fire above the arc of the propeller. Lack of an effective synchronizer was the reason for the awkward mounting. Salmson is holding an automatic pistol, although whether it is for air-to-air use or personal protection in case of being brought down is not known.

Left: The Fokker D.II was basically a biplane development of the earlier *Eindeckers*. It was powered by the same 100 hp Oberursel U.0 used in the E.II and E.III and, like those fighters, mounted a single machine gun. The D.II offered little improvement over the E.II and E.III and was quickly superseded by the Halberstadt and especially the Albatros fighters.

Right: This example of the Fokker D.I has a large fin fitted to a number of aircraft for improved stability. Powered by a 120 hp Mercedes D.II, it mounted a single machine gun. It was the water-cooled engine equivalent to the rotary-powered D.II and was not competitive with the Nieuport fighters. Like the D.II it soon disappeared from the front.

Nieuport 10, Macchi-built.

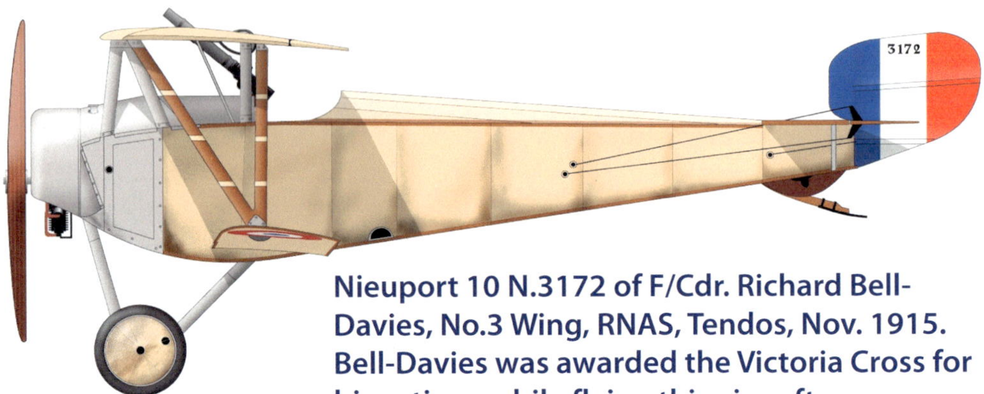

Nieuport 10 N.3172 of F/Cdr. Richard Bell-Davies, No.3 Wing, RNAS, Tendos, Nov. 1915. Bell-Davies was awarded the Victoria Cross for his actions while flying this aircraft.

Nieuport 11 N.576 of Jean Navarre, N.67, Verdun, 1916.

Nieuport 11 N.571 of Sgt. Elliot Cowdin, N.65, March, 1916.

Nieuport 11 N.608 of Lt. Henry Lecour Grandmaison, N.62, 1916.

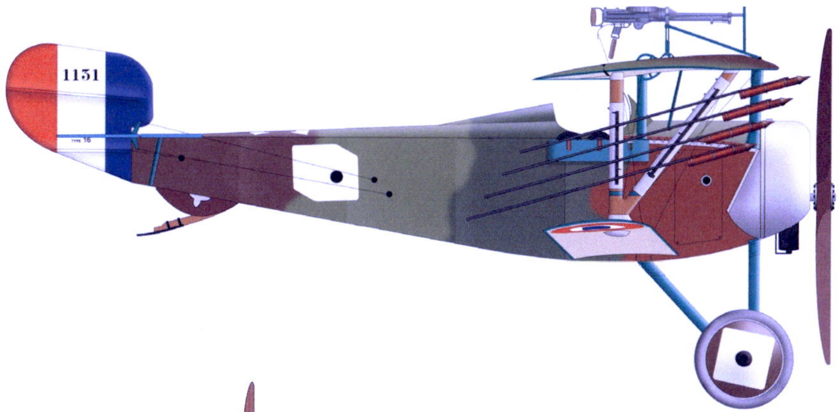

Nieuport 16 N.1131 of Cpl. Charles Johnson, N.124, Behonne Aerodrome, July 1916. Later flown by Sgt. Norman Prince for balloon busting.

Nieuport 20 A6740 of No.45 Sqdn, RFC, April–May 1917.

DH.2 #5985 flown by James McCudden.

DH.2 #5964 flown by Major Lanoe Hawker, No.24 Sqdn, RFC.

FE.8 #6397 flown by 2/Lt. W.B. Hills, A Flt., No.40 Sqdn, RFC, March 1917.

Above: The Fokker D.III was powered by the same 160 hp Oberursel U.III used in the E.IV and, like the E.IV, mounted two machine guns. Like the other Fokker fighters of this time, the airfoil in the center of the upper wing was scooped out to improve the pilot's forward visibility, a feature that reduced lift and created more drag. This D.III, 352/16, was used by ace Oswald Boelcke to down a number of British aircraft while waiting for Albatros fighters to be delivered to *Jasta* 2.

Right: The Fokker D.IV was powered by the 160 hp Mercedes D.III and mounted two machine guns. The D.IV was mediocre at best and few were built; the engine was needed for the far superior Albatros fighters. The failure of this series of Fokker biplane fighters brought Fokker to its lowest fortunes of the war.

Right: The Fokker D.V was the final fighter in this series and reverted to the 100 hp Oberursel U.I and single gun as used by the D.II and earlier E.II and E.III. A delightful aircraft to fly, with excellent maneuverability and handling, its performance was inadequate and most of the D.V fighters built were used for training.

Above: N.20 of 45 Sqdn., RFC mounted a synchronized pilot's gun.

Right: The Nieuport 20 was a development of the N.12 produced for the RFC in 1916; only 21 were built. Powered by a 110 hp Le Rhône, the N.20 had a fixed Vickers for the pilot and the typical flexible Lewis for the observer. A few with the 120 hp Le Rhône were designated N.20bis. Like the earlier N.12, the N.20 was not liked by its crews.

Above: The uninspired Pfalz D 4 was essentially a biplane version of its earlier E.V. Only a few prototypes were built.

Left: *Ltn*. Kissenbeth in his Fokker D.II high over the front.

Above: Navarre in his distinctively-painted Nieuport 11 depicted in a French postcard. The "Sentinel of Verdun" was now a national hero amidst the horrible, depressing slaughter in the trenches.

Jean Navarre was the most famous of the early French aces. Although a few other Allied pilots became aces before Navarre, on 19 May 1916 he was the first Allied ace to reach 10 victories just two months behind Oswald Boelcke (12 March) and Max Immelmann (13 March). Two of the most famous French aces were not far behind; Charles Nungesser scored his 10th victory on 22 June 1916 and George Guynemer scored his 10th victory on 16 July 1916. Both were later to become even more famous than Navarre, who was put out of action on 17 June 1916, being wounded that day shortly after scoring his 12th victory (on that date Guynemer had 9 victories and Nungesser had 8).

Navarre was the prototype for the legend of the fighter pilot. Other early pilots had already become famous, but Navarre was the first 'bad boy' fighter ace. A rebellious youth, his behavior was incorrigible before he became a fighter pilot, much less an ace. He bluffed his way into the *Service Aeronatique* (French Air Service) at the beginning of the war when flight training had been suspended. Prior to his first sanctioned solo flight, Navarre hand-started his aircraft himself, then jumped in and took off without authorization. At his first escadrille, the non-flying commander ordered no acrobatic flying; Navarre would often put on an impromptu air show over the unit's airfield. The commander also ordered all members of the unit to remain on the airfield at all times. Navarre bribed the commander's chauffeur to hide him in the trunk of the commander's car when he went into town in the evenings! The commander denied Navarre's repeated requests to join a more aggressive unit, but one morning the commander found Navarre in town when Navarre prodded the sleeping 'chauffeur' to wake him up – and it turned out it was the commander! Who immediately transferred him! When Navarre received orders to return to his new unit, M.S.12, he 'struggled' to get his Morane Parasol into the air, finally just managing to become airborne before hitting a group of watching officers, who had to dive into the mud in full uniform to avoid being hit by his airplane. Before he could reach his new airfield, his commanding

UN DES ROIS DE L'AIR EN PLEIN VOL : L'AVIATEUR NAVARRE

Celui-là est un " as parmi les as ", c'est-à-dire un des rois de l'air. Voici qu'il a livré plus de quarante combats et abattu quatorze appareils ennemis. Décoré de la Légion d'honneur, de la médaille militaire, de la croix de guerre où les palmes ne se comptent plus, comme un de ses chefs, après un combat victorieux, s'excusait de ne pouvoir lui donner d'autres récompenses, puisqu'il les avait toutes, Navarre trouva cette réponse : " Je mériterai mieux celles que je porte.

Above: The "Sentinel of Verdun" is publicized here in a popular magazine. Navarre is shown at bottom right with his twin brother, Pierre, in the background. Pierre was killed in a flying accident on 15 November 1916 at N.69.

Above: Navarre in his distinctively-painted Nieuport 11 over Verdun.

LE MIROIR

PUBLICATION HEBDOMADAIRE, 18, Rue d'Enghien, PARIS

Le MIROIR paie n'importe quel prix les documents photographiques relatifs à la guerre, présentant un intérêt particulier.

L'ADJUDANT NAVARRE QUI A ABATTU TROIS AVIONS DEVANT VERDUN

Above: Navarre helped create the image of the "bad boy" fighter ace in a number of ways. Here he wears a lady's silk stocking on his head as a way to keep warm. How he got the stocking formed part of the legend…

48 SAN FRANCISCO CHRONICLE, SUNDAY, MAY 28, 1916

Thrilling Duels Fought 4000 Yards Above Earth; Exploits of W

"ACE OF AIR" IS TITLE PEOPLE GIVE NAVARRE

Popular French Airman Fights Alone, Luring Enemy's Craft to Dizzy Heights, Then Looping and Coming Up in Rear to Bombard or Run Them Down

Air Duelists Like Knights to Rescue

THE war in the air is becoming specialized.

Dueling "aces" are recognized to be indespensable.

The recent great battle over Alsace resulted from a French attack on the railroad junction of Muehlhausen and the Aviation Park of Habsheim, center of the German aeroplanes bombarding Belfort. Twenty-three French bombarding machines participated, led by Captain H——, whom the Germans have come to recognize as "Monsieur X——, the Corsair of the Air," from the repeated raids of his squadrons. The attack was a success; yet it is considered, in high French quarters, as proving absolutely that bombarding squadrons must be convoyed by "chasing avions." The dueling "aces" are like knights to the rescue!

By STERLING HEILIG

PARIS, May 27.—"There's the Ace of the Air!"

In the Cafe de la Paix, the maitre d'hotel Michelet indicated a mere boy lunching opposite, alone and conscientiously finishing off with fruit salad and Genois jelly cake.

"It's Navarre!" whispered Michelet.

... they tell such stories? Flying alone in a two-seat "parasol" he got lost in a fog, and came down in the German lines. The machine was intact; but he was surrounded and disarmed. A German officer had an idea. Strapping the young Frenchman securely to the pilot's seat, he jumped in be-...

Jean Navarre, the Ace of the Air, a daring French aviator, and a remarkable photograph, showing him and his warplane in flight above the clouds, on the day he destroyed the fifth enemy aeroplane in his record of 14. The photograph was taken by another French airman.

after all. Those Americans are great!"

Here was my chance to "get" the hero—in Paris, alone, after a good lunch, and well-disposed to Americans for their care of his brother.

Jean and Pierre Navarre are twins. ...

alone, in one-seat "chasing aeros," manipulating rudders and machine-gun single handed, and both are "aces." Pierre Navarre, flying only six months, had piled up a remarkable record when he got these two ...

plane machine gun, in the left shoulder; nasty wound, threatening amputation.

Naturally, Pierre stopped, while Jean, already tie with Georges Guynemer, continued. Later, Guynemer himself was laid up at the Astoria Hotel Hospital with a similar but less dangerous shoulder wound, and Jean Navarre, untouched and fortunate, continued to make victories, and became the Ace with a big "A" for the present.

I sent word by Michelet—would he see me?

Jean Navarre would see me later. He must hasten to visit his brother at the Neuilly Hospital.

Immediately I had an inspiration—the American Hospital! Out there, an "ace" was lying on a bed of pain. Who could criticise Pierre Navarre, flat on his back, for remembering past adventures? If Jean, the big Ace, should be with him, how could he refrain from cheering up the invalid?

TOLD AROUND SICK BED

Here are authentic stories. Some were told by Pierre Navarre to myself. Some were told by Jean Navarre to Georges Prade and a "Matin" man whose name I have missed. All are told as round a sick bed of a wounded "ace." They are mere boys, remember.

"I was discussing with two enemies in a Fokker," a slender, weakly looking blonde boy of 21 was saying. "A broadside of their machine gun put mine out of action. Impossible to shoot a shot. There was nothing left for me to do but to run."

"Of course," murmured the outsiders sympathetically. "Did you get away?"

"No. I was very angry! I ran in—and cut their tail off with a slap of my screw propeller. The screw resisted all right, but it 'deglingued' my motor, and I dived almost 300 yards. I thought I was a goner, but caught myself up in time, and had the satisfaction to see my two Boches dash down to earth alone. I know it was imprudent, but it was the only way to get them."

Jean Navarre, at 5 P. M. the other day, was surrounded by five Fokkers, inside the German lines.

LOOPS TO ENEMY'S REAR

"The Fokkers were far enough away," he says, "at right, left, in front and behind me. I made sure that my 'baby' was in good condition and began to do tricks to amuse them. Suddenly, as by a signal, the five Fokkers dashed straight at me. I waited. I made three little loops, and found myself behind two Fokkers. Naturally, I shot at them—and the others dared not shoot at me for fear of potting their pals. It was extremely rigolo. The whole secret is never to hurry and never let fly your ...

Above: Navarre's fame spread far beyond his native France; the May 28, 1916 issue of the *San Francisco Chronicle* carried his story, complete with photographs, on its front page. Even at this early stage of air combat fighter pilots are already talked about as "Knights"; see the headline "Air Duelists Like Knights to Rescue."

officer had already been notified of the incident by telephone!

Flying to Amiens during a rainy day when flying had been officially cancelled, Navarre and a companion from his unit wined and dined, then flew over the city dropping flowers. Noting some young women waving frantically at them, they flew back to a small airfield and returned to entertain the ladies.

Navarre compensated for his total lack of military discipline with exceptional performance in the air. As celebrated pilot and commander Jean de Rose put it, "With Navarre, I am always being taken by surprise. At the moment when we are going to sign an order of arrest, we are obliged to transform it into a citation."

Navarre, and the pilots like him, had no time for the traditional military emphasis on rigid discipline and over-emphasis on tradition. These men were brave enough to fly at a time when flying itself was very dangerous, let alone combat. And they were disciplined when it counted, in the air when flying against their nation's enemies in combat.

It was at this time that a French newspaper coined the phrase *"l'as de notre aviation"*, meaning 'the ace of our aviation'. The ace was the highest card in the deck, and was used here to denote the finest, most successful aviators. As air combat expanded, the term 'ace' began to be associated with a pilot who had brought down at least five enemy aircraft in combat.

Between his late-night exploits in Paris and his constant 6–8 hours of daily patrolling over Verdun, Navarre exhausted himself. His wounds of 17 June were serious but not fatal, and he recovered physically. His emotional wounds were a different matter. Suffering from what would now be called

La Guerre Aérienne Illustrée

REVUE HEBDOMADAIRE PARAISSANT LE JEUDI

Rédacteur en chef : Jacques MORTANE

ADMINISTRATION ET RÉDACTION : 50, rue de Provence, 50, PARIS. — Téléphone : Bergère 39-61, 39-62

LA MITRAILLEUSE D'UNE DE SES DERNIÈRES VICTIMES

Le capitaine Guynemer était l'un de ceux qui ont abattu le plus d'avions dans nos lignes. Il tenait toujours à aller rendre les honneurs à ses victimes. Guynemer allait auprès de ses Boches afin de prendre leurs armes pour voir s'il ne découvrirait pas une invention, un dispositif qu'il pourrait adopter à son armement.

Above: Navarre's press coverage was not unique. Here ace George Guynemer is featured in a front-page article.

PTSD (Post-Traumatic Stress Disorder), his behavior became out of control and he was transferred to an airman's convalescent home in the country. When problems continued, he was transferred back to his unit but crashed on his first test flight after fainting due to his wounds reopening in flight. Then he received news that his twin brother, Pierre, had been killed in a flying accident, and Navarre went into a deep depression. Close friends were killed in combat or flying accidents. On a trip to Paris he drove his car on the sidewalks and actually ran over a policeman trying to stop him; he was arrested. The doctors agreed Navarre was emotionally unstable and not responsible for his actions, and he was sent home to his family. Postwar he was given a job as a test pilot. On 19 June 1919, while practicing for a flight under the Arc de Triomphe that he decided to do in conjunction with a grand victory parade, he hit telephone wires and crashed to his death. He is buried alongside his twin brother in Landes, France.

Right: Navarre's recovery was also a matter of national interest; here the caption reads "Our wounded ace Jean Navarre".

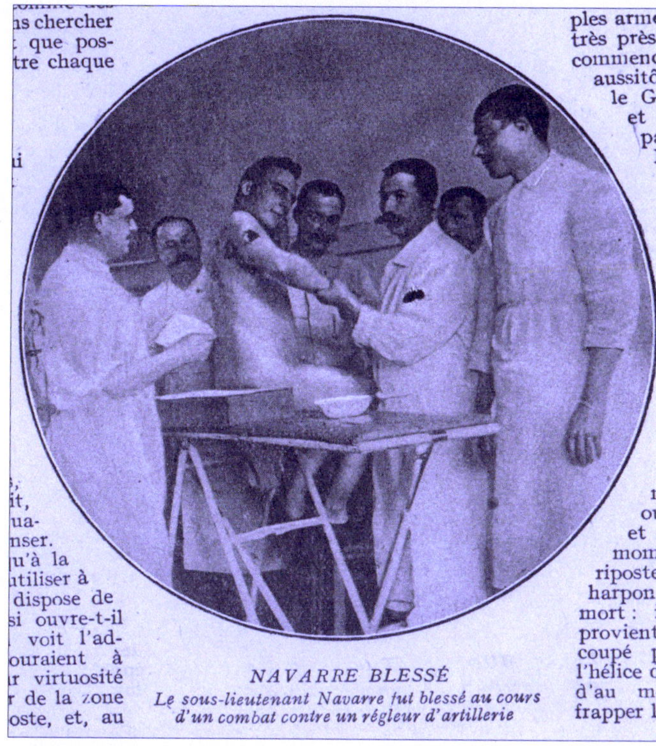

NAVARRE BLESSÉ

Le sous-lieutenant Navarre fut blessé au cours d'un combat contre un régleur d'artillerie

Above: Navarre's condition after being wounded was covered in the French press. No privacy here!

Fokker D.I 159/16,
unit unknown, 1916

Fokker D.II of *Kest* 4b, Freiburg
Aerodrome, 1916

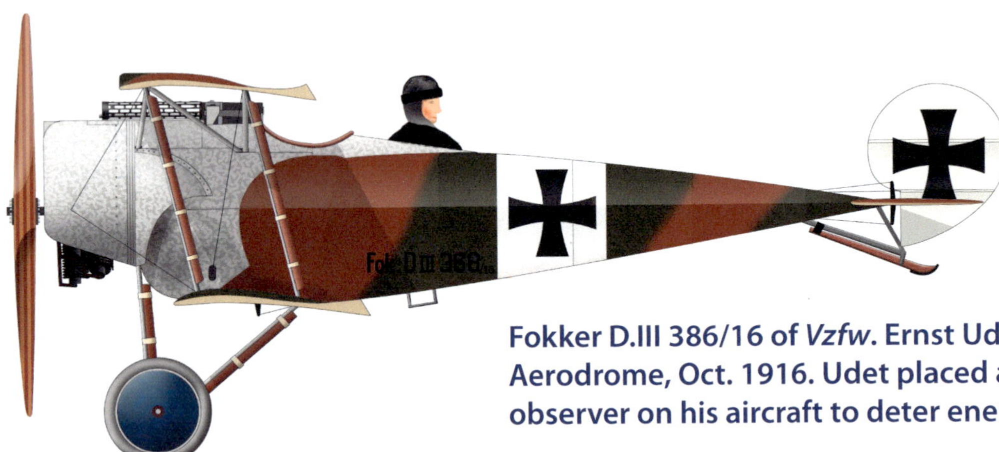

Fokker D.III 386/16 of *Vzfw*. Ernst Udet, *Jasta* 15, Habscheim Aerodrome, Oct. 1916. Udet placed a tinplate cutout of an observer on his aircraft to deter enemy fighter attacks.

Fokker D.V 2X82/16 "X" of an unknown
naval base defense flight, 1916/1917

Above: The first prototype of the Russian MB (Mosca-Bystritskiy), built in 1915, was powered by a 50 hp Gnôme rotary. S. Gilber and F. Mosca are in the cockpit and Capt. Bystritskiy is facing the camera. Ten similar aircraft were ordered as 'fighters' with no information about armament. In 1916 the single-seat MBbis was tested; it had an 80 hp Gnôme and an upward-firing Lewis gun without synchronizer. More than 20 were built and used operationally in 1917.

Above: Powered by a 100 hp Gnôme rotary, the Sage Type 2 was an unusual two-seat British fighter design that first flew in August 1916. Designed before synchronizing gear was available, the gunner, located behind the pilot in an enclosed cabin, stood up through the upper wing and had an excellent field of fire. By the time the Sage Type 2 was flown synchronizers had become available and development was abandoned. Only a single prototype was built.

Third Generation Fighters

Above: The Halberstadt D.II was the first German biplane fighter, and first third-generation German fighter, to reach the front. With one synchronized machine gun and 120 hp Mercedes D.II engine, it was far superior to the Fokker *Eindeckers* it replaced, offering better performance, better maneuverability, and much greater structural strength.

Their loss of air superiority at Verdun forced the Germans out of their lethargy and they started searching for an answer. An effective response had to be both technological and tactical, because the Allies were superior in both areas. Even as the Fokkers were being driven from the sky over Verdun, new German fighters were being designed and tested. Fokker delivered a new family of biplane fighters, the D.I/D.II/D.III that, while better than his monoplanes, still were not as good as the Nieuport. The Halberstadt company delivered an excellent new series of biplane fighters that were strong, maneuverable, and delightful to fly. The Halberstadt fighters brought the Germans to technical parity with the Allies, but more was needed.

That more was delivered by the powerful Albatros firm, the largest airplane manufacturer in Germany. The sleek new Albatros D.I and D.II combined good streamlining with a powerful 160 hp Mercedes six-cylinder engine, giving them enough power to carry two synchronized machine guns. The result was a fighter that was faster and more heavily armed than the Nieuports and DH.2's of the Allies. Hedging their bets, the German authorities also decided to copy the French Nieuport, which clearly had more development potential than the pusher DH.2. Passing some captured Nieuports around key manufacturers resulted in the Siemens-Schukert D.I, essentially a copy of the Nieuport powered by an innovative Siemens-Halske counter-rotary engine. It was used in small numbers at the Front, but by the time it reached combat it was out-classed by newer designs.

With the technology advantage once more in their favor due to the Albatros fighters, the Germans needed to close the tactical and organizational gaps. Finally they decided they must emulate the fighter squadrons of the Allies, and Boelcke was re-called to the Front to form one of the first Jadgstaffeln (literally hunting squadron), abbreviated as Jasta. During his travels Boelcke met many German airman, and while on the Russian Front before

Third Generation Fighters	At Front
Albatros D.I &I D.II	8/16–9/17
Albatros D.III	1/17–9/18
Albatros D.V & D.Va	6/17–Armistice
de Havilland DH.5	5/17–3/18
Halberstadt D.II/D.III/D.V	6/16–7/17
Morane-Saulnier AC	~1/17–5/17
Nieuport 17, 23, 24, 25, 27	6/16–5/18
Pfalz D.III & D.IIIa	9/17–Armistice
Roland C.II	3/16–5/17
Roland D.I	10/16–7/17
Roland D.II & D.IIa	2/17–12/17
Roland D.III	2/18–5/18
Sopwith 1½ Strutter	4/16–10/17
Sopwith Pup	9/16–9/17
Sopwith Triplane	2/17–11/17
SSW D.I	4/17–??/17

Third generation fighters were the first to arrive at the front that were designed in response to the experience of fighter combat. These aircraft had one or more fixed guns for the pilot, and most of these were synchronized guns. They used more powerful engines and had stronger airframes than their predecessors to withstand high-G maneuvering combat.

Fourth generation fighters were the second generation of fighters to arrive at the front designed based on lessons learned from combat experience. They were refinements of the previous generation technology, but what these fighters did *not* have was innovative new engines or structural and aerodynamic designs that characterized fifth generation fighters.

For Allied fighters, the innovations were all about the engines, either advanced rotary engines like the Sopwith Snipe's Bentley B.R.2 or powerful, water-cooled V-8 or V-12 engines like the Hispano-Suiza V-8 or Rolls-Royce V-12. For the German fighters, the structural and aerodynamic advances came first with Fokker's innovative cantilever wings, followed by the Sh.III counter-rotary engine and the BMW.IIIa over-compressed engine, then the Mercedes D.IIIaü over-compressed engine.

Sixth-generation fighters, just entering service at the end of the war, had all-metal structures. There were two of these, the Junkers D.I single-seater and its two-seat derivative the Junkers CL.I. By far the most structurally advanced fighter was the Zeppelin D.I biplane with all-metal, semi-monocoque construction. This advanced aircraft previewed the structural technology used for decades after it appeared, but was not placed in production.

In his book Hallion discusses five fighter generations and calls the Sopwith Camel a fourth-generation fighter, apparently due to its combat success. But did the Camel offer any real innovations compared to the preceding Sopwith Pup or Triplane? My answer is 'no'. Its structure was the same technology; the only real difference was that for the first time Sopwith engineers performed a proper stress analysis on the structure. So the Camel was more robust than the Pup and Triplane, but not fundamentally different. In fact, the much later Sopwith Snipe used the same structural and aerodynamic technology to its detriment, and only its advanced rotary engine qualified it as a later-generation fighter. Some may point to the Camel's armament of two synchronized Vickers, but a few Triplanes also carried that armament.

To categorize the fighters usefully, six generations are needed; my third through fifth generations further expand on Hallion's third and fourth. The Camel was basically a refinement of earlier Sopwith fighters based on combat experience, and no new technology was introduced. The Camel could carry four 20-pound Cooper bombs, which made it more effective for ground attack, expanding its versatility. So, while the Camel was effective enough in combat to serve until the Armistice, it was a fourth generation fighter that had no technical features qualifying it as a fifth-generation fighter.

Similarly, the Hispano-Suiza-powered Spad 7, which first reached the front in small numbers starting in August 1916, was the first fourth-generation fighter to see combat. Initially its Hisso engine only produced 140 hp, less than the 160 hp Mercedes in-line six-cylinder in the Albatros. More powerful Hisso's soon arrived and most Spad 7 fighters had 180 hp engines, with some late production aircraft having 200 hp engines, making it a transitional fighter between fourth and fifth generations. The 180 hp Hisso gave the Spad 7 better performance than the Albatros, and paved the way for the fifth generation Spad 13.

The Bristol F.2A/B, SE.5a, and Spad 13 were the earliest fifth-generation fighters and all served successfully until the Armistice.

Above: The Halberstadt fighters had no fixed tail surfaces yet were very robust despite their delicate appearance.

Above: The Nieuport 17, an enlarged development of the Nieuport 16 that eliminated the nose-heaviness of that type, was powered by the same 110 hp Le Rhône used in the N.16. Reaching the front in June 1916, it was the first French third-generation fighter. Performance was better than earlier Nieuports but inferior to the Albatros fighters that were soon introduced. This aircraft was flown by famous ace George Guynemer of N.3, later Spa.3. The *cône de pénétration* was not a spinner and did not rotate, but was fixed in position. This aircraft carried one synchronized gun, although armament varied; some aircraft had a fixed over-wing gun, others had both, which reduced performance.

returning home, he met a number of eager young pilots and told them he was returning to the Western Front to form a *Jasta* with pilots he chose. The pilots quickly realized he was evaluating them. The next morning he appeared at the hut of two of them and asked "Want to come to the Somme with me?" *Ltn.* Erwin Böhme and *Ulanen Leutnant* Manfred von Richthofen enthusiastically accepted.

Boelcke returned to the Western Front in late August and formally established *Jadgdstaffel* 2 (also called *Jasta* 2) by the end of the month. The first fighters, an Albatros D.I and two Fokker D.III's, arrived on 1 September. In early September Boelcke trained his pilots thoroughly in combat flying. He developed a set of eight rules, known as Boelcke's Dicta, for successful fighter combat. These rules, listed in an adjacent table, still apply today. In parallel with training his new pilots Boelcke also flew missions by himself and scored more victories. On Sunday, 17 September, came the graduation exercise. Boelcke led five of his pupils on patrol and they soon spotted a British formation of BE.2c's and FE.2b's below them. The British dropped their bombs just before the Albatros fighters attacked. The Germans caught the British by surprise and a violent dogfight erupted. When it was over, four FE.2b's and two BE.2c's had fallen; the six Albatros fighters of *Jasta* 2 returned home without loss.

The next month the Germans re-organized their entire air service. As early as March 1916 *Oblt.* Hermann von der Lieth-Thomsen, the dynamic German *Feldflugchef* (Chief of Field Aviation), had proposed that all German aviation organizations should be combined under one central authority. While the German navy did not like this idea, the initial failure of German aviation over the Somme eliminated doubts, and on 8 October the Kaiser issued a decree placing all army aviation units under the authority of the commanding officer of the air services. *Generalleutnant* Ernst von Hoeppener, a professional cavalry officer, was appointed to this position, *Kommandierenden General der*

Above: The Albatros D.I, here D.435/16, was the first of the Albatros fighters. With its streamlined fuselage, twin synchronized machine guns, and 160 hp Mercedes engine, it immediately established technical superiority over Allied aircraft at the front. The Albatros fighters were the most important of the early third generation fighters.

Below: The Albatros D.II, here D.386/16, was the second Albatros fighter to reach combat. It was an improved D.I; the cabane struts were redesigned for better forward view and the upper wing was lowered closer to the pilot's line of sight, also improving forward view. The rest of the aircraft remained the same, as did its performance.

Left: The Sopwith 1½ Strutter, or Strutter for short, was the first British two-seater with a synchronized gun for the pilot, making it more effective in combat. This aircraft, A993 "A-6" of 43 Sqdn., RFC, was shot down on 28 April 1917 while being flown by 2/Lt. C.M. Reese and 2A.M. A. Moult over Lens-Neuvireuil.

Right: The Sopwith Scout was quickly nicknamed the 'Pup' due to its resemblance to is larger predecessor, the 1½ Strutter. Most Pups carried a fixed, synchronized Vickers. The Pup was the first Sopwith single-seat fighter to be built in large numbers. Powered by an 80 hp rotary, it was a maneuverable aircraft that was delightful to fly. The Pup could hold its own with the Albatros but was normally on the defensive due to the Albatros's superior performance and armament.

Luftstreitkräfte (*Kogenluft*), General in command of the German Army Air Service. Thomsen was appointed as his chief of staff, and *Major* Siegert remained at *Idflieg* (*Inspektion der Fliegertruppen* — Inspectorate of Aviation Troops). These three men ran the *Luftstreitkräfte* (the Imperial German Air Service) until the end of the war. Like the RFC, the *Luftstreitkräfte* was part of the army, not a fully-independent air service. In von Hoeppner's words, the Germans now realized "numbers, leadership, fighting spirit, and the technical excellence of fighter airplanes themselves would increasingly determine the outcome of the struggle for air superiority."

The Germans had learned their lessons; they had combined the latest tactics, training, and organization with the best available technology. The result was German air superiority until the late Spring of 1917, when new Allied fighters arrived. The German fighter pilots became significantly more aggressive and started scoring heavily against Allied airmen. The improved training instigated by Boelcke and adopted by other *Jasta* leaders also came at a time when British training standards had reached a low point. In September the RFC lost 170 airmen in combat, more than two thirds in the last half of the month following introduction of the new Jasta tactics. British casualties continued to rise in October, and the life expectancy of a new RFC pilot at the Front dropped to three weeks.

Tragically, although he played a key role in resurrecting German air superiority, Boelcke was not to enjoy it for long. On 28 October 1916, with his victory score standing at 40, the first ace on either side to reach this number, Boelcke was killed as the result of a mid-air collision during combat. Taking off with Böhme, his closest friend, von Richthofen, and three other pilots to intercept British airplanes that had just crossed the lines, they came upon two

Boelcke's Dicta

1. Seek an advantage before attacking. If possible, keep the sun behind you.
2. Once you've started an attack, always carry it through.
3. Fire only at close range and only when the opponent is properly in your sights.
4. Always keep your eye on your opponent, and don't let yourself be deceived by ruses.
5. Whenever you attack, it's important to go at your opponent from behind.
6. If your opponent attacks you from above, don't try to evade him but fly right at him.
7. When you are over enemy territory, never forget your own line of retreat.
8. For the *Staffel*: attack on principle in a group of four or six; but when the fight breaks down into individual combats, several shouldn't go after a single opponent.

In early July 1916, Chief of Field Aviation Hermann Thomsen met with Oswald Boelcke at German Supreme Headquarters at Charleville. "At my request he drew up the following summary of the principles that should govern every air fight; briefly composed and simply expressed, they were also to serve as a source of success for the younger scouts. These principles established by Boelcke remained in force until the end of the war." Indeed, they were only slightly revised and employed by fighter pilots until the age of jets and avionics. Because of Boelcke's Dicta (see above) and the fact that he was the first pilot to create and promote fighter team tactics in the air, he is often recognized as the father of fighter aviation.

Above: Flying reproduction of the Sopwith Triplane, the successor to the Sopwith Pup. The Triplane followed the basic construction of the Pup aside from its new triplane wing cellule, and its 110 hp or 130 hp Clerget rotary gave it improved performance. Nearly all Triplanes had the single fixed Vickers of the Pup, but a few late-production Triplanes had two guns. The Sopwith Triplane was the only British aircraft in early 1917 that could cope with the Albatros D.III, and in face slightly out-performed the Albatros. The Sopwith's triplane configuation was thought by *Idflieg* to be key to its performance and this started Germany's 'triplane craze' of mid-1917 to the end of the year.

Left: A Nieuport attacks a German observation balloon with Le Prieur rockets. The rockets had very short range and 8–10 were carried on the interplane struts. Although the rockets were very inaccurate, one hit would burn a balloon.

DH.2's of No.24 Squadron, RFC. During the attack that followed, Böhme and Boelcke both closed on the same DH.2 when another flew across their path. Böhme and Boelcke maneuvered abruptly to avoid a collision with it, and in doing so lost sight of one another because their wings blocked their view. As they recovered from this maneuver, Boelcke's upper left wing lightly touched Böhme's undercarriage, breaking it. Böhme quickly recovered and watched in growing concern as Boelcke's Albatros glided down in an ever-steepening dive, crashing near a German artillery battery. Unrestrained by a safety harness, Boelcke was killed in a crash that he likely would have survived otherwise.

Boelcke's death shocked the German fighter community and the nation. An RFC airplane flew over Boelcke's elaborate funeral and dropped a laurel wreath with the inscription "To the memory of Captain Boelcke, our brave and chivalrous foe. From the British Royal Flying Corps." By imperial decree the Kaiser renamed *Jasta* 2 as *Jasta* Boelcke, with Boelcke's life to serve as a model for all aspiring fighter pilots. More than just a great fighter pilot, Boelcke was the first great fighter tactician, and his Dicta became the official tactical air doctrine of the German fighter forces. He also set a wonderful example by intensely training his pilots before combat. In addition, he was one of the uncommon examples of chivalry in an air war that has been overly romanticized, perhaps partly in comparison to the ghastly horror on the ground. Early in his flying career Boelcke rescued a drowning French boy and the villagers wrote the French government requesting he be given a medal for his heroic action! Not surprisingly, no French medal was forthcoming, although he did receive a German lifesaving medal for his deed.

On 23 November von Richthofen avenged Boelcke's death by shooting down Major Lanoe Hawker, VC, who was CO of No. 24 Squadron that had figured in Boelcke's death. Hawker was an excellent pilot and early ace who was awarded the Victoria Cross, Britain's highest honor, after shooting down two German aircraft while flying the Bristol Scout, with its obliquely-firing armament. It was the first VC awarded for air combat between airplanes. In September 1915 Hawker was posted to command newly-formed No. 24 Squadron, destined to be the RFC's first fighter squadron. After forming up and receiving its DH.2's, the squadron flew to St. Omer in France on 8 February 1916. Hawker's combat tactics were simpler than Boelcke's; a board was erected in the squadron area that said: "Tactical Orders by O.C. No.24 Squadron, Royal Flying Corps: Attack everything." It was signed by Hawker. On the fateful day, Hawker, now with 12 victories, ended up in single combat with von Richthofen in his Albatros D.II. The DH.2 was more maneuverable but the Albatros had every other advantage, and during the fight lasting some minutes, Richthofen had the initiative and Hawker was on the defensive. During the fight the prevailing westerly wind was

Nieuport 17 of *Cne*. Georges Matton, OC, N.48, late 1916.

Nieuport 23 of *Adj*. Edmund Thieffry, *5eme Escadrille*, Belgian Air Service 1917. Thieffry scored 10 victories.

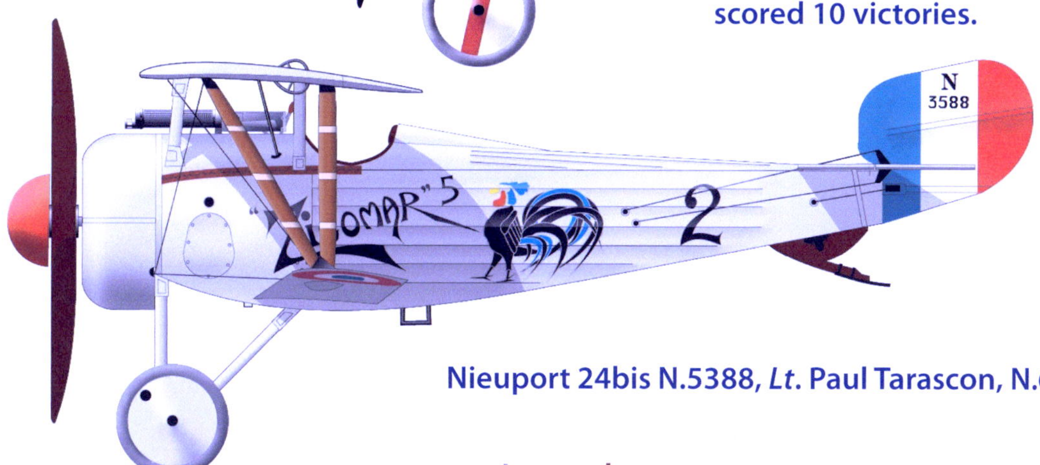

Nieuport 24bis N.5388, *Lt*. Paul Tarascon, N.62

Nieuport 27 of Capt. Wendell Rogers, No.1 Sqdn., RFC. On 12 December 1917 Rogers shot down a Gotha bomber with this N.27 for his 7th of an eventual 9 victories.

Albatros D.I D.391/16 of Ltn. Karl-Heinrich Büttner, *Jasta* Boelcke, November 1916.

Albatros D.II of *Ltn*. Manfred von Richthofen, *Jasta* 2, September 1916.

Albatros D.III of *Jasta* 36, Le Chatelet Aerodrome, Spring 1917.

Albatros D.V of *Ltn*. Wolf, *Jasta* 5, July 1917.

Right: Impressed by the Nieuport's performance and maneuverability, *Idflieg* sent data and captured aircraft to major German manufacturers and encouraged them to explore the Nieuport's sesquiplane wing cellule. Albatros responded by fitting a similar wing cellule to their D.II fighter, creating the D.III. The new wing had less drag and the D.III was faster and climbed better than the D.I & D.II. The D.III further enhanced the technical superiority of the Albatros fighters – but at a price. The new wing cellule was not as robust due to the single-spar lower wing, and lower wing failures during high-G combat started to occur.

Above: The lighter, more streamlined Albatros D.V reached the front in May 1917. The lighter structure was also weaker, and wing failures became a serious problem. Moreover, performance was little improved. For most of 1917 into early 1918, the Albatros D.V/D.Va was the most numerous German fighter, and was out-classed by 4th-generation Allied fighters like the SE.5a and Spad 7 & 13. This D.Va is preserved in the Australian War Memorial, Canberra.

The death of Oswald Boelcke in a mid-air collision during air combat was a shock to Germany. At that point Boelcke had 40 victories, the first ace to reach that score, and was the world's leading ace and a German national hero. As the adjacent articles from the *New York Times* indicate, his fame spread far beyond Germany.

Photographs of his funeral on the facing page show that his death was commemorated as if he were royalty. In fact, he was royalty, a prince of warriors in this new realm of war in the air, a rank he had won by courage, skill, and accomplishments. In attendance at his funeral were dignitaries such as Crown Prince Rupprecht of Bavaria, *Gen.* Fritz von Below (the Kaiser's repressentative), *Gen.* Sixt von Arnim, and *Gen.* Wolf Marschall von Altengottern. When Boelcke's coffin was loaded on a train back to Germany, an honor guard fired three volleys. Once in Germany, a second funeral service was held at which *Gen.* Moritz von Lynker, the Kaiser's representative, delivered the eulogy. When that service was over and the coffin was carried out, it seemed like the entire population of the city of Dessau were waiting in mourning attire. Aircraft flew overhead and the Duke of Anhalt awaited Boelcke's coffin at the cemetary. At Boelcke's final resting place,

BOELKE REPORTED KILLED.

Aeroplane Like German Champion's Is Brought Down by Ribiere.

LONDON, June 17.—Captain Boelke, the champion German aviator, who received an autograph letter from the Emperor recently complimenting him on his success in bringing down French aeroplanes, is believed to have been killed in an aerial combat with the French aviator, Roger Ribiere.

A Fokker painted yellow and of the type known to have been piloted by Boelke, who also wore a large yellow muffler, was shot down by Ribiere two days ago between the German and French trenches near Verdun.

Captain Boelke and his colleague, Lieutenant Immelmann, have been mentioned more times in the German official statement than any other aviators in the aerial branch of the Teuton army. Up to May 21 Captain Boelke had accounted for eighteen French aeroplanes. The last hostile machine brought down by him was vanquished in an aerial battle over Dead Man Hill in the region of Verdun.

A month ago the German airman was promoted from the rank of Lieutenant to that of Captain by Emperor William in acknowledgment of his achievements.

BOELKE STILL FIGHTING, AMERICAN AIRMEN SAY

Victor Chapman Thinks He Fought Noted German Aviator on Saturday.

Special Cable to THE NEW YORK TIMES.

PARIS, June 20.—According to American airmen there is no truth in the report that Boelke, the noted German aviator, was recently killed in a duel with a French airman. A letter just received here from Victor Chapman, a member of the American squadron, says:

"Last Saturday I had the narrowest escape to date. Coming home late from a reconnoissance alone. I was attacked from behind by a Fokker swooping from a great height. This, I think, was Boelke, as that is his invariable method of attack. A regular hail of bullets all round the machine riddled it. The controls were so damaged that only by holding them with the hand was I able to make a landing. I was lucky to get off with a light scalp wound, insufficient to cause me to leave the front."

Mr. Chapman added that his assailant was flying a black machine which coincides with American accounts of Boelke. In contradiction of a Matin story which described him as wearing yellow headgear and flying a yellow machine, Americans say that Boelke's aeroplane is black with a huge skull painted under each wing. This, they think, is a deliberate attempt to unnerve the opponents upon whom he swoops.

Mr. Chapman was also in a perilous position Friday evening, when he was surrounded by a group of enemies, but Rockwell came to his rescue and the two eluded the Germans without injury.

German air raiders bombarded Bar-le-Duc on Friday and Saturday. On Friday the Americans would in ordinary circumstances have returned before the raid, which occurred about dusk, but all prolonged their work in the hope of accomplishing something to mark the day, which was the anniversary of the great Artois battle last year, when the first regiment of the Foreign Legion suffered severely and four Americans, Weeks, Kelly, Fike, and Hall, were killed.

Above & Left: These are the two articles in *The New York Times* that initially reported Boelcke's death but then printed Victor Chapman's testimony that Boelcke was still very much alive.

Above: Boelcke's coffin (center right) rests on a horse-drawn gun carriage in front of Cambrai Cathedral before the procession to the rail station. Prof. Max Bölcke and his wife (center) stand on the cathedral steps wearing black mourning clothes.

Right: Manfred von Richthofen (far right), carrying Boelcke's *Ordenskissen*, leads the procession into Cambrai rail station.

Oberstleutnant Thomsen, Chief of Staff of the *Luftstreitkräfte* (German Air Force) delivered a stirring speech later published everywhere in Germany that declared that all German airmen must aspire to the solemn vow "I will be a Boelcke!"

By proclamation of the Kaiser, Boelcke's unit, *Jasta* 2, was given the honor name *Jasta* Boelcke, and it was decreed that fresh laurel wreaths from the Air Force would be placed on his grave on each anniversary of his death.

Above: The Albatros D.III was based on the late-production D.II airframe with a new wing cellule inspired by the Nieuport sesquiplanes. Reports of the Nieuports' maneuverability and performance inspired *Idflieg* to send captured Nieuports around to different German manufacturers for inspection. Albatros based the wings of their D.III on the Nieuport design, producing the famous "V-Strutter" Albatros fighters. The new, higher aspect ratio wings provided better performance and downward vision than the previous wing, but inherited the weakness of the Nieuport's single-spar lower wing. Unknown to *Idflieg*, Nieuports had already experienced failures of their lower wing, and the faster, heavier Albatros D.III suffered even more. The problem became acute when the Albatros D.V, a lightened D.III design, reached combat in May 1917.

pushing the two combatants further into German lines. Finally, Hawker had to break off combat and run for British lines; Richthofen shot him down and Hawker was killed. Both pilots were experienced aces, and Richthofen owed this victory, his 11th out of an eventual 80 that made him the leading ace of the war, to the Albatros' superiority.

New Allied Fighters

While the Germans introduced the superior Albatros and other new biplane fighters, the Allies also continued developing new fighters. The 80 hp Nieuport 11 was given a more powerful 110 hp engine to create the nose-heavy Nieuport 16; an enlarged and more balanced airframe resulted in the Nieuport 17 that arrived in June. In most Nieuport 17 fighters a synchronized gun replaced the over-wing gun.

The French SPAD company's first fighter series was the unusual SA series of 'pulpit' two-seat fighters. The SA series were tractor biplanes with a gunner's nacelle mounted in front of the engine and propeller to give a wide field of fire. In addition to its additional weight, the nacelle limited propeller efficiency and impeded the flow of cooling air to the engine, limiting the power of the engine that could be fitted. The concept was a failure and only about a hundred were built. Only one French SPAD SA succeeded in downing a German airplane, although a few more victories were scored by the SA in Russian service. The story could have ended there with failure. However, the SA series, while flawed in concept, had a robust airframe, and in one of the great aviation success stories of the war, the failed SA series was developed into the highly successful SPAD 7 that was one of the best fighters of the war. By eliminating the troublesome gunner's nacelle, installing a synchronized machine gun, and replacing the under-powered rotary engine with the innovative 150 hp Hispano-Suiza V-8, SPAD transformed the SA failure into the SPAD 7 success. All subsequent wartime SPAD aircraft were developed from the SPAD 7, and SPAD fighters were supplied to Britain, Italy, Belgium, Russia, and the USA.

The SPAD 7 was a very strong airplane and its 150 hp Hispano-Suiza engine, later boosted to 180 hp, gave it excellent performance. Arriving at the Front the same month as the Albatros, its main shortcoming was its single machine gun compared to the two guns of the Albatros. However, the SPAD also had teething problems with its engine and radiator that initially limited its combat effectiveness and, for a time, slowed the numbers that reached service; later it was produced in great quantity.

While French industry was focusing on Nieuports and the new SPAD 7, the British developed the Sopwith 'Pup' fighter, so called because it looked like a smaller version of the earlier Strutter. The Pup was delightful to fly and a good match for the

Above: The LFG Roland C.II was a break-through design for Roland. The exceptional streamlining for the time – it entered service in March 1916 – was developed in conjunction with wind-tunnel testing and full-size testing on a high-speed train. To reduce drag the upper wing was built directly into the fuselage without benefit of cabane struts. The C.II was the smallest, fastest German two-seater when it reached the front, and was faster than most Allied fighters. Initially armed with a single flexible gun for the observer, the second and subsequent production batches included a fixed, synchronized gun for the pilot as shown above. The speed of the C.II coupled with its effective armament made it the first German two-seater that was effective as a two-seat fighter. The relatively high-performance C.II proved a handful for many service pilots and its maneuverability and flying qualities were mediocre. The narrow gap between the wings resulted in airflow interference between them, reducing lift and increasing drag. The view upward was excellent but the view downward and forward was limited, a problem that contributed to numerous landing accidents. *Lt.* Seitz was the pilot of this C.II. The Roland D-series single-seat fighters were derived from the C.II.

Above: A camouflaged Roland C.II of *Flieger-Abteilung* 62 with tactical number '7' painted on its fin and a shark mouth.

Despite Albert Ball's youthful appearance, he was the United Kingdom's leading ace at the time of his death in May, 1917, with 44 victories at 20 years old. Ball was the first fighter pilot to capture the British public's imagination and he was successful at a particularly difficult time for the RFC, when the German Air Service was in the ascendency.

Born 4 August 1896, Ball had a happy childhood, part of which he spent in his own shed behind the family home where he would build things like small boats and work on small engines. His success at school was average, but he continued learning about all things mechanical. Ball left school at age 17 and with his father's help began his own business.

He enlisted immediately upon the onset of the First World War, and was assigned to train recruits. Wanting to get to the action in France, he transferred to a motorcycle company. However, he remained in England as a dispatch rider. Frustrated, Ball turned to aviation as a means to be posted to France. Ball quickly came to enjoy flying and found his tinkering with engines and electrical items came in very handy. At the front he would usually help or direct his ground crew with the maintenance of his aircraft. He received his RFC wings on January 29, 1916, and was assigned to 13 Squadron, RFC in early February; there he flew BE2 aircraft on observation missions. His first combat occurred on 29 March when Ball dove at a German aircraft only to be attacked in turn by another German aircraft and forced to defend himself. Interestingly, Ball wrote home after this, his first encounter with the enemy: "I like this job, but nerves do not last long, and you soon want a rest." This letter came after only six weeks of front-line duty!

The squadron had acquired a small, single-seat Bristol Scout, and Ball found his calling by using this aircraft to chase after Germans. His desire to engage the enemy did not go unnoticed and in

Above: Albert Ball, VC. His youth is apparent; he was only 20 when killed in May 1917. Ball lasted longer than most lone-wolf hunters; by the time of his death combats between flights and squadrons of fighters were common.

Facing Page: Ball holding the red spinner he added to his Nieuport to increase its speed.

Below: Ball and his Nieuport 16 A201 without spinner.

May 1916 Ball received a transfer to 11 Sqdn., RFC that operated as a fighter squadron. Within his first month there, Ball scored eight victories!

Ball's favorite method of attack was to fly below his opponent and then utilize his upper-wing Lewis gun to fire into the German aircraft from below in its blind spot. He flew as many missions as he could, and with his approach of attacking the enemy regardless of the situation, Ball rapidly burned out emotionally. He wrote to his parents:"My work is rather a nerve pull, so I think it best for me to forget all about it after it is over."

As the Somme battle raged, Ball went full steam every day, but began to feel fatigued and approached his commanding officer to request a brief leave. Instead, he was sent to 8 Sqdn., RFC to fly the slow, vulnerable BE2 aircraft again! Apparently, when Ball's C.O. forwarded his request to the commanding general, the general took umbrage with the request from such a junior officer and decided to not only refuse the leave, but to teach the youngster a lesson.

Ball would spend nearly a month flying the boring but dangerous missions that were given to the observation squadrons. Then, in early August Ball received a letter requesting his return to 11 Sqdn. On his arrival, he found a brand new Nieuport scout that was to be his personal aircraft: he viewed both events as birthday presents.

From this time on, including a transfer to 60 Sqdn., Ball flew multiple sorties every day weather allowed. He was aggressive and attacked large enemy formations, diving into them and finding the aircraft that seemed to be piloted poorly. He would attack that aircraft until he shot it down. He would then try to escape, stating that there was safety in numbers – in this case the enemy would have trouble attacking him without endangering a comrade.

On several occasions, Ball fired a salvo of Le Prieur rockets into the German formation to break it up, then attack an aircraft. He constantly worked on his aircraft and his engine to enhance its performance, and added a spinner painted bright red in the hope that it would add to the aircraft's speed.

Finally, in October 1916 Albert Ball was sent home on a long leave and then re-assigned to Home Establishment duties. Six months after arriving in France, he had returned to England as a true hero. He was promoted to the rank of Captain, he was awarded medals directly by the King at the palace, including the Victoria Cross, and he received many honors from his hometown of Nottingham. He was the first fighter ace of the Royal Flying Corps who was given so much fame and publicity. It had been an unwritten rule that specific names would not be given to the news media from the RFC, but Ball was proving to be too much of a star to keep out of the public eye.

Ball was appointed a flight commander in a new squadron being created, No. 56, which was to be equipped with the latest British fighter, the SE5.

Ball returned to France with his new squadron, but such was his fame that he was allowed to bring a Nieuport to use as he did not like the assigned SE5. Although he flew both aircraft, he much preferred the nimble Nieuport because it was a better fit for his fighting style.

After No. 56 Squadron arrived in France in late April 1917, Ball continued to dive into the enemy formations regardless of the odds, and often had to nurse a heavily damaged fighter back to his airfield.

On 6th May 1917, Ball wrote to his father: "I do get tired of always having to kill and am beginning to feel like a murderer. Shall be so pleased when I have finished." That same night he wrote to his fiancée: "Oh, won't it be nice when all this beastly killing is over and we can just enjoy ourselves and not hurt anyone."

The next day, Ball was seen to emerge from cloud cover in an inverted position and dive into the ground, dying from the crash.

Albert Ball's achievements, courage, and devotion to duty continue to inspire British fighter pilots.

Above: The DH.5 was an attempt to give the pilot an unimpeded field of view upward and forward; to achieve that, the wings were staggered backward. Unfortunately, this reduced visibility aft, a detriment in air to air combat. Armament was a single synchronized Vickers machine gun.

Albatros in a dogfight, but its 80 hp engine only enabled it to carry a single synchronized machine gun. The first of a long line of successful Sopwith fighters, the Pup was clearly superior to the DH.2 pusher in all respects, and was stronger than similarly-powered Nieuports.

Bloody April

Despite the newer Allied fighters being introduced, momentum was with the German *Jastas*. In late 1916, the powerful, two-gun Albatros was the best fighter at the front, and the new *Jastas* made the most of it. And this time, the Germans did not rest on their technology laurels; with the Albatros D.II superior to Allied fighters and strength building rapidly, the Germans introduced the Albatros D.III with even better performance. The D.III adapted the Nieuport sesquiplane (one and a half wing) wing cellule to the basic Albatros design, improving its downard visibility, speed, and climb rate. However, the single-spar lower wing of the Nieuport was not robust, a weakness exacerbated by the greater weight and speed of the Albatros D.III, which subjected the wing to greater stresses. This would lead to problems.

Meanwhile, Trenchard was alarmed about the appearance of superior German fighters, their use of better tactics, and their growing numbers. He communicated his concern as early as September, and in December 1916 visited London to urge that his fighter strength be increased dramatically and soon. The only effective way this could be done was to introduce a significant number of new fighter types with better performance. Unfortunately, production and training difficulties delayed delivery of most of these better airplanes until late Spring 1917.

In January 1917 von Richthofen was finally awarded the *Pour le Mérite*, known familiarly as the Blue Max, Germany's highest award, after his 16th victory and given command of *Jasta* 11. Until Richthofen arrived, *Jasta* 11 had not claimed a single victory. Richthofen immediately solved that problem; the day he arrived at *Jasta* 11 he shot down an FE.8 fighter, a pusher design much like the DH.2 and obsolete when it reached the front. The next day he downed an FE.2b. Richthofen also instructed his pilots in Boelcke's Dicta and that training, together with his dynamic leadership by example, transformed the unit.

In March 1917 the weather improved significantly, enabling increased air activity. Air fighting became much more intense, and British losses rose to 120 aircraft, almost half over the British lines. Trenchard now faced a dilemma; his losses were high in the face of German technical superiority, but the imminent Arras offensive required extensive reconnaissance well behind German lines. British hours flown per casualty fell from 186 hours in October 1916 to a mere 101 hours in March 1917. At the same time, the demand for more and deeper

Above: Powered by a 110 hp Le Rhône, the DH.5 was generally inferior to Albatros fighters when it was introduced to combat. Its maneuverability was good but performance was mediocre and worse than mediocre above 10,000 feet. The DH.5 was not popular with pilots due to its unusual configuration and modest performance and it was replaced by superior British fighters as soon as practical. A robust aircraft, it was useful for low-level ground attack.

reconnaissance missions increased as the date for the Arras offensive approached. Despite increasing casualties, Trenchard had no choice but to order the essential reconnaissance flights.

An example of the difficult month of March was on the 24th, when 70 Squadron dispatched six Sopwith Strutters on a reconnaissance mission. A dozen Albatros fighters bounced them, shooting down two and riddling the rest, which barely made it home. The next day 70 Squadron attempted to complete the mission with another group of six Strutters. One survived because it turned back early with engine trouble; the other five were shot down over the German lines by Albatros fighters of *Jastas* 5 and 6. In eleven months the Strutter had gone from triumph over the *Eindeckers* to obsolescence against the Albatros. So great were the casualties that the RFC had to recruit gunners from the infantry. Frank Courtney, a flight commander in 70 Squadron, related that a volunteer infantry gunner arrived at the squadron and during his briefing, asked why the squadron was known as the 'Suicide Club;' he was posted missing the next day.

The numbers favored the British, who had 754 aircraft along their front, including 385 fighters, arrayed against 264 German airplanes, 114 of which were fighters. Some of the British fighters were Nieuports, Sopwith Pups, and Spads, but most of them were the old DH.2 and FE.2b. Trenchard decided to go on the offensive on 4 April, five days before the Arras ground offensive was to begin. Things did not go well. On 5 April a formation of six new Bristol F.2a two-seat fighters from No.48 Squadron, the first of their kind at the Front, flew their first mission over the lines lead by Capt. Leefe Robinson, VC. Robinson had been awarded the VC for shooting down a German airship at night over Britain in 1916. Unfortunately, Robinson had no experience in combat against other fighters, and he insisted on following defensive tactics whereby the pilot was supposed to maneuver the airplane to give the gunner the best opportunity for a shot. These tactics were appropriate for a typical reconnaissance two-seater, but the new Bristol F.2a was designed as a fighter. As luck would have it, the six Bristol Fighters ran into five Albatros fighters from *Jasta* 11 lead by von Richthofen. The Bristols flew like any other group of two-seaters and *Jasta* 11 treated them that way. The Albatros fighters shot down four of the Bristols, including Robinson's, and the other two returned riddled with bullets. Robinson survived as a PoW, but died two months after the war in the worldwide flu epidemic.

This disaster was a major setback to Trenchard's hopes for the new fighter. Fortunately, it was a case of inexperience and poor tactics overcoming excellent technology; this would be overcome as the RFC gradually learned to exploit the Bristol's

Above: Nieuport 23 of the Belgian Air Service, which flew Nieuport 11s, 16s, and 23s, but no Nieuport 17s. Differences between the Nieuport 17 and 23 were slight, the main difference being use of a new interrupter gear in the Nieuport 23 that required the Vickers gun to be offset a few centimeters to the right of centerline.

performance as a fighter. When properly used it became the best two-seat fighter of the war. The F.2a model was powered by a 190 hp Rolls Royce Falcon V-12, and the F.2b had a more powerful, 275 hp version of the Falcon that gave it even better speed and climb. With good maneuverability for its size, excellent performance, and good armament – a synchronized Vickers for the pilot and flexible Lewis for the gunner – it could be flown offensively like a single-seat fighter, but with the added protection of the rear gunner. Unfortunately, this learning curve took too long to avoid the heavy casualties of Bloody April.

The intense fighting that opened April continued without respite for the entire month. The twin-gunned Albatros wrecked havoc with the single-gunned Allied fighters and obsolescent reconnaissance planes. Gordon Taylor, an RFC Pup pilot, had an opportunity to fly a captured Albatros D.II a few weeks after Bloody April; he thought the Albatros "a war machine, a weapon of ruthless efficiency". In contrast, he viewed the Pup as suitable for sport flying after the war. While the Pup could out-maneuver the Albatros, the Albatros had twice the firepower and twice the engine power, and being faster and able to out-climb and out-dive the Pup, usually had the initiative. The Albatros could also break off combat at will if a Pup got the tactical advantage. Amongst the RFC fighters the Nieuports took especially heavy losses; in two weeks 60 Squadron lost 13 pilots.

Within the first four days of the aerial offensive, the RFC lost 75 airplanes in combat and a further 56 airplanes in accidents. Squadron Commander Sholto Douglas recalled that his unit, 43 Squadron, took more than 100 percent casualties during April while flying the Strutter; 32 pilots and gunners were authorized the squadron, and 35 were lost. Replacements arrived just in time to be shot down, and "six or seven" of the original 32 squadron members survived April. So many escorts had to be provided, at least five for a reconnaissance plane, that the nominal RFC numerical superiority was nullified.

British losses were severe; 245 airplanes known lost directly as a result of combat, 211 aircrew killed, missing, or died of wounds, 108 taken PoW, and 116 wounded in action. In comparison, the RFC lost 499 aircrew in the four and a half months between 1 July and 22 November 1916, the period of the Somme. British aircrew never gave up, but by the end of April 1917 RFC morale was at its nadir.

When war broke out, George Guynemer immediately volunteered for duty in the air service, but was refused three times on account of his frail health until he was eventually accepted as a mechanic. His persistence ultimately lead to gaining his pilot's license and an assignment to M.S.3.

Guynemer was the first fighter pilot to score three victories in a single day, he helped design the cannon-armed Spad 12, and he used it successfully in combat. He also worked with Kodak and utilized a camera to record the first ever gun camera images.

Guynemer earned the respect and affection of his fellow fighter pilots because of his audacity, skills, and incredible courage. Guynemer was shot down on six occasions, one time by a French anti-aircraft unit. But the true measure of George Guynemer is seen in his unsurpassed devotion to duty and love of his country. After scoring his 50th victory, he was on a brief leave at home when his father suggested that he should not fly in combat anymore, he had done enough already. To which, Guynemer replied:" If one has not given everything, one has given nothing."

On 11 September 1917 Guynemer took off on a mission and never returned. He apparently was brought down in combat and his body was identified by German soldiers who retrieved his identification card, but he had fallen in an area with numerous artillery barrages so neither his aircraft nor his body were recovered. At the time he had 53 victories and was the leading French ace.

The following was written in *L'Illustration* on 6th of October 1917: "He was neither seen nor heard as he fell; his body and his machine were never found. Where has he gone? By what wings did he manage to glide into immortality? Nobody knows: nothing is known. He ascended and never came back, that is all. Perhaps our descendants will say: He flew so high that he could not come down again." Indeed, at the time French schoolchildren were taught that.

However, no praise could better describe the sense of the Guynemer legend than the plaque that is to be found in the Pantheon:

"Captain Guynemer, commander of Squadron No.3, died on the field of honor September 11, 1917. A hero of legendary power, he fell in the wide heaven of glory, after three years of hard fighting. He will long remain the purest symbol of the qualities of the race: indomitable in tenacity, enthusiastic in energy, sublime in courage. Animated with inextinguishable faith in victory, he bequeaths to the French soldier the imperishable remembrance which will exalt the spirit of sacrifice and the most noble emulation."

No other French pilot would ever gain the fame, the adulation, and affection of his nation as George Guynemer did during the war. Even today, the *Armee de L'Air* (France's modern air force) holds an annual parade on the 11th of September, where his last posthumous citation is read aloud to all ranks. George Guynemer's ancestors, who were French knights, must be proud.

Above: George Guynemer by his Nieuport 10, the first of six fighters he named *Vieux Charles* (Old Charles) in tribute to a comrade from N.3, Charles Bonnard, who had been transferred to the Macedonian Front.

The *Lafayette Escadrille* holds an especially romantic appeal for American aviation enthusiasts. On 21 August 1914, William Thaw and 41 other Americans then in France enlisted in the French Foreign Legion. In early October, while in combat with the Second Regiment, Thaw related to Paul Rockwell his hope of forming "a squadron of American volunteers" to fly for France. On 24 December 1914 Thaw joined *Escadrille* D.6 as a *soldat-mitrailleur,* thus becoming the first American to serve in a French squadron at the front.

Having convinced the French *Aeronatique Militaire* that he was a qualified pilot, on 1 February 1915 Thaw was sent to the aviation school at Saint-Cyr, On 15 March Thaw received his *Brevet Militaire* #714 on the Caudron G.2, the first American to be so-licensed by the French government. Thaw served with distinction as a pilot and was promoted to Sergeant on 18 May 1915. He was cited and awarded France's *Croix de Guerre* with palm on three occasions.

About the time Thaw was serving at the front, Norman Prince, a Massachusetts aviator, had completed his French aviation training and had arrived at the front on 20 May 1915, joining V.B.108. Prince had gone to France for the purpose of forming an all-American volunteer squadron of aviators to fly for the French Air Service. To this end, he solicited the support of the wealthy American colony residing in France and also contacted ministers within the French government.

Prince, Thaw, and Elliot Cowdin, an American pilot serving in Prince's squadron, pressed their efforts at forming an all-American squadron. Their dreams were realized on 16 April 1916 when the "*Escadrille Americane*" was established. Officially designated N.124, the squadron and its seven American pilots began operations at Luxeuil-les-Bains, southeast of Paris. The pilots flew their first combat patrols in Nieuport 11 fighters on 13 May. The *escadrille* was commanded by French officers, and eventually 38 Americans served with it.

America will still neutral at this time and the Germans protested the provocative name of the unit, which was then changed to *Lafayette Escadrille* in honor of the Marquis de Lafayette, a French hero of the American and French revolutions.

A total of 265 American pilots flew for France as part of the Lafayette Flying Corps; however, the *Lafayette Escadrille* was the only French unit with nearly all American pilots.

During 23 months of combat flying, the squadron's 38 American pilots flew more than 3,000 combat patrols, scoring 39 confirmed victories over enemy aircraft. Eleven *Lafayette Escadrille* pilots, including Norman Prince, were killed.

This was a respectable combat record, but of greater importance, as intended by its founders, was the sympathetic response of Americans to this volunteer unit, which helped bring America into the war against Germany in April 1917. Also important was the combat experience of the pilots, which was put to good use by the USAS when it began operations. Many of the pilots transferred to the USAS and were placed in leadership positions to help the novice units.

Left: Victor Chapman, the first US airman to lose his life in combat. Chapman, flying with the *Lafayette Escadrille*, was flying his Nieuport 16 on June 23, 1916 on a personal errand to deliver a crate of oranges to a hospitalized friend. He noticed a dogfight in which he recognized three of his squadron mates. Chapman joined the fight only to find that the other Nieuports had escaped and he was now alone against five German Fokker *Eindeckers*; Chapman was soon shot down and killed.

Above: James Norman Hall, together with fellow Lafayette Flying Corps pilot Charles Nordhoff, co-authored *Mutiny on the Bounty* and many similar adventures postwar.

Above: The *Lafayette Escadrille* adopted two lion cubs they named Whiskey and Soda as mascots. Here William Thaw holds Whiskey.

Below: The lions soon grew up; here Whiskey is photographed with Raoul Lufberry. With 16 victories Lufberry was the most successful member of the *Lafayette Escadrille*.

Right: Lt. William Thaw, one of the original founders of the *Lafayette Escadrille,* is credited with the idea of using the famous Seminole Indian head as the group's logo. On Feb. 8, 1918, the unit was disbanded and 12 of its members transferred to the USAS as part of the 103rd Aero Squadron.

Above: Raoul Lufberry after transferring to the USAS 94th Aero Squadron as its CO early in 1918. He scored his 17th and last victory with the 94th Aero. While flying a Nieuport 28 on May 19 he attacked a Rumpler reconnaissance aircaft. Its gunner set his Nieuport afire and, with no parachute, he jumped to his death from his burning fighter.

Above: The Morane-Saulnier AC with synchronized gun was the next stage of development after the 'N' and was used at the front in small numbers. The Allies were generally skeptical of the monoplane configuration and Morane-Saulnier pursued monoplane fighter designs longer than other Allied companies. The thin airfoil used on the AC meant extensive bracing was required for sufficient wing strength, and the drag generated by this bracing limited any advantage the monoplane might offer over the biplane. Given the elementary structural technology available, the biplane configuration was lighter and stronger than the monoplane with only a modest drag penalty. The comparative fragility of monoplanes made them unpopular and monoplane designs were generally avoided as air combat intensified.

Sopwith 'Pup' N6205 *BETTY* flown by Flt. Cdr. Joseph Fall of 3 Sqdn., RNAS. Fall scored 3 of his 36 victories flying N6205.

Sopwith Triplane N533 ***Black Maria*** of Flt. Lt. Raymond Collishaw, No.10(N) Sqdn., RNAS, July 1917. Collishaw scored a total of 62 victories.

Right: Nieuport 24 of French *Escadrille* N91. The Nieuport 24 featured re-designed tail surfaces and rounded fuselage. These changes improved its appearance but did not, as hoped, improve its performance. With the Nieuport 17/23 the basic configuration had reached the limit of its development potential. The airframe could not accommodate heavier, more powerful engines and further developments, such as the Nieuport 24, 25, and 27, were no improvement.

Right: The Nieuport 24bis reverted to the same tailplane as the Nieuport 17. Differences between the types included the airfoil section and the 130 hp Le Rhône in the 24bis. The Nieuport 24bis actually entered service before the Nieuport 24, but both were inferior to contemporary German fighters. The Nieuport 25 may have been a one-off, and the Nieuport 27 was nearly identical to the Nieuport 24; the differences were small modifications to the landing gear.

Below: Nieuport 27 of the French *Escadrille* Spa 87. By the time the Nieuport 27 arrived, it was only used because Spad production could not meet demand. Nieuport pilots now had to hope to survive until re-equipment with Spads, and *Lt.* Marin shown here did not make it; he was shot down and killed in combat with *Jasta* 43 on February 5, 1918.

As a youth, Charles Nungesser travelled to South America to seek adventure. Attending a boxing match in Buenos Aires, Nungesser was incensed when a French boxer was knocked out almost immediately by an Argentine boxer, who then began to insult all Frenchmen. Nungesser immediately stepped up to face the Argentine boxer and fought until he prevailed, indicating the type of fighter pilot he would become. He experienced numerous crashes with multiple injuries, and at times he would have to be carried to his aircraft.

At the beginning of the war, Nungesser enlisted in the 2nd Hussars. During the first weeks of the war, he and some other French soldiers ambushed a German staff car, which Nungesser then drove at speed through various outposts and skirmishes to deliver the dispatch cases that had been discovered in the car. For this action, he was invited to lunch at divisional HQ and given the *Medal Militaire* and the staff car itself as a reward for his daring!

Once combat settled into trench warfare, Nungesser joined the air service and by April 1915 was flying a Voison pusher in V.B.106, where he first began using personal markings of a large, black skull and crossbones that were later modified to the black heart with coffin seen in the adjacent photos.

Nungesser soon transferred to fighters. Arriving at

Below: Nungesser in his Nieuport 16 with his macabre personal markings.

Right: Nungesser beside his Nieuport 17; his macabre personal markings had become more elaborate than his earlier skull and crossbones.

his new squadron, he put on a display of aerobatics over the airfield. His commanding officer was not happy, "[if you want to] impress someone with your flying skills, frighten the enemy!" Nungesser stormed out of the office and immediately flew to a German aerodrome where he proceeded to put on a display of aerobatics, which included flying upside down past the airfield's hangars. He returned to his own airfield, informed his commander that he had successfully completed his mission, and was placed on eight days arrest!

Once on lone patrol, he was attacked by a British fighter that was so persistent that Nungesser was forced to shoot it down. After that tragedy, Nungesser painted prominent stripes in the French national colors, red/white/blue, on his aircraft to help prevent a repeat occurrence.

Nungesser's score of 45 victories places him third on the list of French aces, and he was one of the few early French fighter pilots to survive the war.

On May 8, 1927 he took off from Le Bourget in his specially constructed *Oiseau Blanc* (White Bird) in an attempt to be first to fly non-stop across the Atlantic. He was never seen or heard from again.

Right: Nieuport 25 with Nungesser's personal markings. This is the only Nieuport 25 that has been positively identified and it may have been the only Nieuport 25 built.

Above & Below: The aircraft shown here are not captured Nieuport fighters but early and late production SSW D.Is. One of *Idflieg's* poorer decisions was to send captured Nieuport fighters around to German aircraft manufacturers and suggest they try the Nieuport's sesquiplane wing cellule. Albatros adapted the sesquiplane wing to create their D.III and D.V fighters, but SSW and Euler directly copied the Nieuport airframe. At least SSW used their own innovative Sh.I counter-rotary engine instead of a Le Rhône. The 110 hp Sh.I needed a front bearing support, which can be seen on the D.I above. Due to its counter-rotary design, the propeller turned slowly and needed to be of large diameter, so the undercarriage was taller than on the Nieuport. Copying a design is not a good idea; by the time it is brought to production, the original design team will have produced something better. This should have been obvious, but apparently was not. Surprisingly, the SSW D.I reached limited combat service (the one below at *Jasta* 5), but only 95 were built before production was stopped because the aircraft was obsolete by the time it reached service. However, the exercise was not completely wasted; SSW went on to design their own original fighters using the enlarged, more powerful Sh.III engine, and these aircraft were exceptional performers.

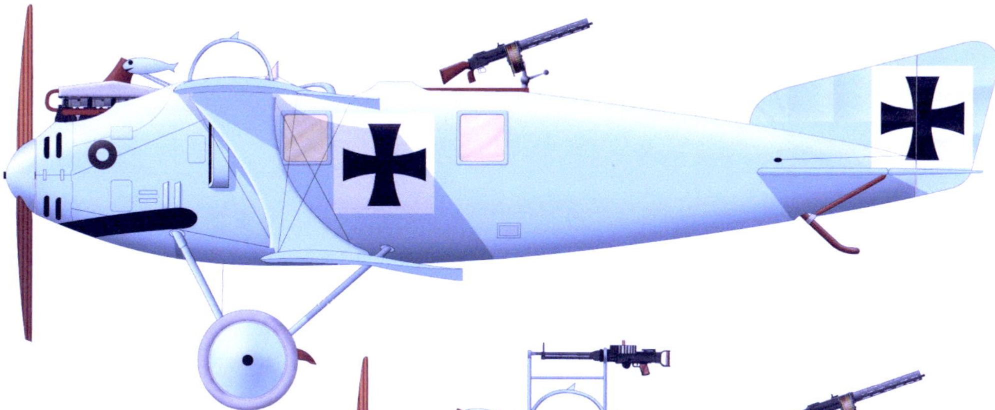

Roland C.II flown by *Hptm*. Eduard *Ritter* von Schleich of *Schusta* 28 in original configuration.

Roland C.II flown by *Lt*. Otto Czermack of *Schusta* 28 after a forward-firing gun was fitted in a crude mount.

Roland C.II with shark mouth of *Flieger-Abteilung* 62.

Roland C.II of *Schusta* 27.

SSW D.I 3761/16 of *Jasta* 5.

Roland D.I of *Jasta* 5.

Roland D.IIa of *Jasta* 25 in Macedonia. Most likely this was the aircraft flown by 19-victory ace *Vzfw*. Gerhard Fieseler.

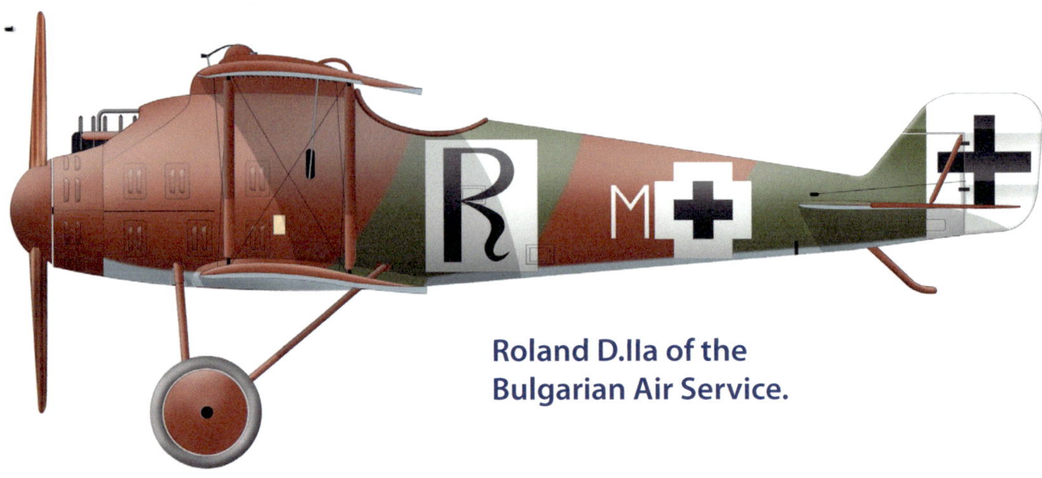

Roland D.IIa of the Bulgarian Air Service.

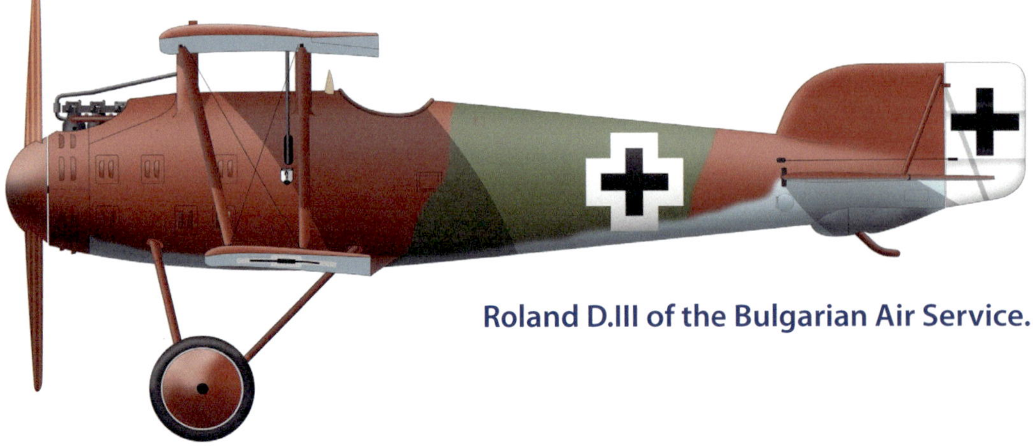

Roland D.III of the Bulgarian Air Service.

Above: The production LFG Roland D.I shown here at Adlershof was also built under license by Pfalz as the Pfalz D.I. The crosses are of normal proportions, which indicates this may be a Pfalz-built aircraft; Roland-built aircraft had thicker crosses. Power was a 160 hp Mercedes and two synchronized guns were mounted.

Above Right: The Roland D.II had a redesigned center section to improve the pilot's forward field of view and an airfoil radiator replaced the 'ear' radiators used in the D.I. The D.II and D.IIa, powered by the 180 hp Argus, were the main production models of the early Roland fighters.

Right: The Roland D.III was another re-design to improve the pilot's view. The lower wings were of narrower chord; both 160 hp Mercedes and 180 hp Argus engines were used.

The Pfalz D.III used the *Wickelrumpf* technique of fuselage construction that Pfalz had learned from building LFG Roland D.I & D.II fighters under license. This entailed wrapping thin, narrow strips of plywood around a mold (see below) and gluing them together into a strong, light, monocoque shell. During the late stages of the production process the fuselage was wrapped by a layer of fabric. Minimum framing was required. The technique facilitated streamlined shapes and conserved scarce resources by using abundant plywood. It also resisted battle damage because the entire shell carried the load, leaving no weak spots in the structure that could be disabled by a single bullet. Although labor intensive, this construction proved so successful that it was used by all known subsequent Pfalz designs, including the Pfalz D.VIII, whose fuselage shells are shown ready for assembly at left. Ironically, starting with the D.IV Roland moved to a less labor-intensive method for fuselage construction similar to clinker-planked boat hulls, with the result it was called *Klinkerrumpf* (clinker body).

Captured Pfalz D.III 4184/17 stripped bare for inspection. The light fuselage framework supported an outer plywood shell, created by the *Wickelrumpf* technique. The barrel of one of the buried machine guns is just visible.

Right: Pfalz D.III 4011/17 of *Lt*. Fritz Höhn in *Jasta* 21, winter of 1917–1918, with his teddy bear mascot. This D.III wore the *Jasta* 21 unit marking of a vertical black and white line just aft of the cockpit. The top surface of the wings and the rest of the aircraft markings were Höhn's individual markings. Höhn scored 21 confirmed victories, including 10 balloons. He later became acting commander of *Jasta* 60, then commander of *Jasta* 41, where he scored three victories in three days, but was then killed in action.

Right: Pfalz D.III 4011/17 of *Jasta* 21, flown by *Lt*. Fritz Höhn, winter of 1917–18. The machines of this unit also carried small personal numbers in black, in this case marked beneath the serial number on the fuselage. The combination of horizontal and diagonal stripes were an attempt to make the aircraft more difficult for ground gunners to hit, and were applied as Höhn was beginning his career as a balloon buster.

Right: The elegant Pfalz D.VI appeared in January 1917, well before the better-known Pfalz D.III. It was not placed in production, perhaps due to limited supply of the Castrol lubricant required by its 110 hp Oberursel Ur.II rotary, and by the First Fighter Competition in January 1918 was slower than the D.VII and D.VIII powered by the 160 hp Siemens-Halske Sh.III.

This Page: The FE.9 two-seat fighter/reconnaissance aircraft was apparently intended as a replacement for the FE.2 series. It was powered by a 200 hp Hispano-Suiza but by the time it appeared in early 1917 the pusher configuration was already obsolete and only three were built. Despite its advanced Hispano-Suiza engine, the FE.9 used the obsolete pusher configuration and as a result the author considers it a third generation fighter. The FE.9 was developed into the NE.1 night fighter; six prototypes of the NE.1 were built but its pusher configuration was outdated.

Sopwith 1½ Strutter A8226 of No.45 Sqdn., RFC.

DH.5 A9507 *Elsa* of No.64 Sqdn., RFC.

Sopwith Pup A7309 "A" flown by Lt. P.G. Taylor, No.66 Sqdn., RFC, Estree Blanche Aerodrome, September 1917.

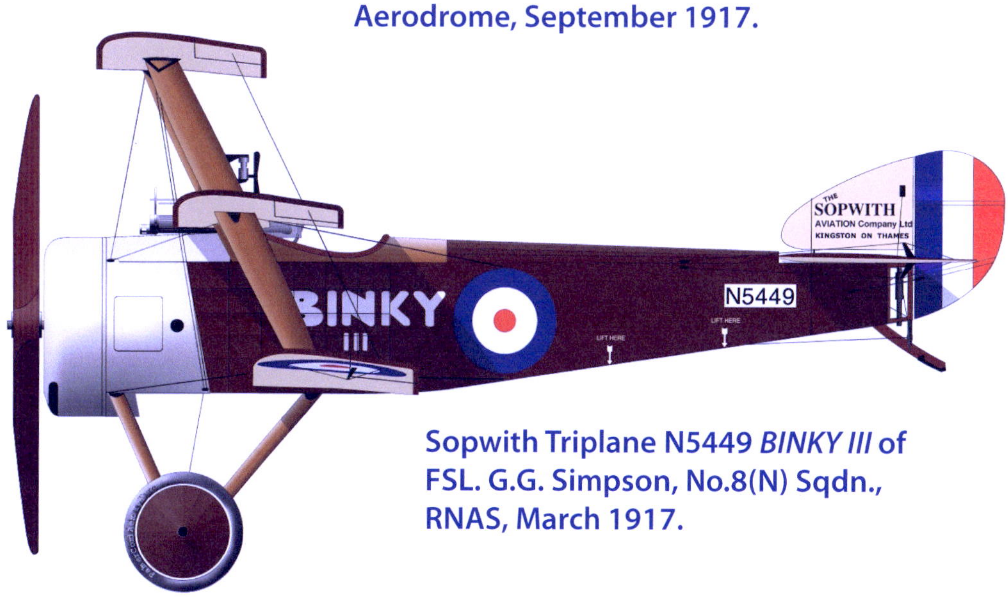

Sopwith Triplane N5449 *BINKY III* of FSL. G.G. Simpson, No.8(N) Sqdn., RNAS, March 1917.

Above: Unlike most E-types that had rotary engines, the NFW E.II had a 160 hp Mercedes D.III, giving it much better performance. Although a pleasing design with fairly good speed, only one aircraft was built.

The National Flugzeug-Werke (NFW) at Leipzig was primarily involved in aircraft repair and flight training. However, NFW built a couple of fighter prototypes. The NFW E.I was a typical *Eindecker* except for its wing, but the NFW E.II was an interesting example of a third-generation German fighter prototype. Built in 1917 to the design of *Dipl. Ing.* Heinrich, it featured a combination of first-generation *Eindecker* configuration with a powerful engine and greatly improved wing design.

In contrast to the Fokker and Pfalz *Eindeckers* that used wire-braced wings and wing-warping for roll control, the NFW E.II had plywood-covered cantilever wings with minimal wire bracing and was fitted with ailerons for roll control. The two wing spars were continuous through the fuselage for greater strength and the pilot sat between them.

Unlike most *Eindecker* fighters, the NFW E.II was powered by a water-cooled engine, in this case a 160 hp Mercedes D.III. The more streamlined design and greater power resulted in much improved performance compared to earlier *Eindeckers*, with maximum speed improved by 30 km/h, making it faster than Albatros fighters, and the climb rate more than doubled despite its greater weight.

NFW E.II Specifications

Engine:	160 hp Mercedes D.III	
Wing:	Span	12.00 m
	Wing Area	17.00 m^2
General:	Empty Weight	558 kg
	Loaded Weight	768 kg
Maximum Speed:		186 kmh
Climb:	2,900 m	6.3 min.

Although the NFW E.II displayed good performance, nothing is known about its handling qualities or maneuverability, both critical aspects of a fighter. In any case, for unknown reasons this rare, little-known fighter was limited to a single prototype.

The NFW E.II was the most advanced German monoplane fighter design before the advent of the Fokker monoplane prototypes of 1918 (V17, V23, V25, V27, V28) with their thick, plywood-covered, cantilever wings. Unfortunately, the NFW E.II lacked the innovative Fokker wing technology.

Above & Below: The NFW E.II had under-wing radiators to cool its 160 hp Mercedes D.III, and a streamlined nose with spinner. The result was much improved speed and climb compared to the earlier Fokker *Eindeckers*. "National", for National Flugzeug-Werke, was painted on the fuselage side, and a turn-over structure protected the pilot.

Fourth Generation Fighters

Above: Restored Spad 7 held in the National Museum of the USAF. The Spad 7 was a great success and was flown in combat by France, Belgium, Britain, Italy, Russia, and the USA. The Spad 7 initially used the 140 hp Hispano-Suzia V-8. Engine power was soon developed to 150 hp, then 180 hp, making the Spad 7 faster than the Albatros. Only one synchronized gun was fitted. Arriving at the front about the same time as the Sopwith Triplane, the Spad did not make the same emotional impression on *Idflieg*, perhaps because it was a biplane. Regardless, the robust Spad airframe had more development potential than the Sopwith Triplane and the more powerful, two-gun Spad 13 became the standard French and American fighter until the end of the war.

Below: Despite their streamlined appearance, the third-generation Albatros fighters were at a disadvantage to the fourth-generation Spad 7 in airframe strength, speed, climb, and dive. The Albatros D.III ruled the skies in April 1917 and the D.V arrived at the front in May. By July Manfred von Richthofen was complaining about the "wretched Albatros." This example is preserved in the Australian War Memorial, Canberra.

FOURTH GENERATION FIGHTERS	AT FRONT
Fokker Dr.I	9/17–Armistice
Halberstadt CL.II	11/17–Armistice
Halberstadt CL.IV	8/18–Armistice
Hannover CL.II	11/17–Armistice
Hannover CL.III	4/18–Armistice
Hannover CL.IIIa	4/18–Armistice
Hanriot HD.1	8/17–Armistice
Morane-Saulnier AI	2/18–5/18
Nieuport 28	5/18–7/18
Pfalz D.XII	6/18–Armistice
Roland D.VIa & D.VIb	6/18–Armistice
SE.5	4/17–8/17
Sopwith Camel	6/17–Armistice
Spad 7	8/16– Armistice

The fourth generation fighters were the third generation of fighters to reach the front after designers had the benefit of feedback from the front about fighter combat. They included the robust Hanriot and twin-gun Sopwith Camel and Nieuport 28, which finally abandoned the out-dated sesquiplane wing cellule. Germany produced their effective two-seat fighters and the Roland D.VI and Pfalz D.XII. Fourth gneration fighters were also the first to incorporate important technical advancements; the early Hispano-Suiza V-8 engines for the Allies and the thick airfoil, cantilever wing for the Fokker Triplane. As such, they were more refined and effective fighters that included important transitional types and some of the most famous fighters of the war.

The Hispano-Suiza-powered Spad 7, which first reached the front in small numbers starting in August 1916, was the first fourth-generation fighter to see combat. The initial production version produced 140 hp, less than the 160 hp Mercedes in the Albatros, and it had a single synchronized machine gun while the Albatros had two guns. Later Spad 7 fighters had 180 hp Hispano-Suiza engines, with some late production aircraft having 200 hp engines, making the Spad 7 a transitional type from fourth to fifth generation.

Similarly, the 150 hp Hisso powered the British SE.5 fighter that entered combat in April 1917 with No.56 Squadron. The SE.5 was another transitional type, soon giving way to the 200 hp SE.5a. Only a small number of SE.5 fighters were produced before production shifted to the SE.5a, but the Spad 7 was produced in great numbers, as was the Sopwith Camel, making these two types the most numerous and important of the fourth-generation fighters.

The most famous fourth-generation German fighter was the Fokker Dr.I, the famous Triplane. The Dr.I introduced Fokker's thick, wood cantilever wing into production, and as such was itself a transitional design to the much more potent Fokker D.VII fifth-generation fighter. The Roland D.VI and Pfalz D.XII single-seaters arrived at the front in mid-1918, but were preceded by the Halberstadt and Hannover two-seat fighters. Consequently, the Halberstadt and Hannover fighters were far more important operationally. In contrast, the tardy Roland D.VI and Pfalz D.XII were only stop-gaps until Fokker D.VII production could replace them in the *Jastas*. The fifth-generation Fokker D.VII was a much superior fighter eagerly awaited by pilots.

Above: The Spad 7 fighter with a single synchronized gun and powered by a Hispano-Suiza V-8 of 150 hp was the first 4th-generation fighter to appear in combat, arriving in August 1916. A second-generation Nieuport 16 is in the background. This Spad 7 was from the *Lafayette Escadrille* and was photographed in April 1917.

Above: Good view of a typical Camel.

Right: The Camel was a versatile fighter and could carry up to four 20# Cooper bombs for ground attack; here Cooper bombs are being loaded on a Camel. The SE.5a was much faster, especially at higher altitudes, and as a result Camels were frequently used for ground attack. This was one reason Camel pilot casualties were so much higher than those for the SE.5a. Of course, another was the Camel's eccentric handling qualities.

The iconic Sopwith Camel was one of the most important and versatile fourth generation fighters.

Left: A typical Camel day fighter, its design role. Camels were also used for night fighting, ground attack, and as shipboard fighters.

Above: Reproduction Sopwith Camel in the markings of William Barker, VC. The Sopwith Camel is rightly one of the most important and famous fighters of the war. The Camel followed the famous Sopwith Triplane in production and improved on the Triplane in two important ways. First, it was much more robust, being the first Sopwith design subjected to a formal engineering analysis of its structure. Second, it mounted two fixed, synchronized guns compared to the single gun standard on the Triplane. Its performance was no better than the Triplane, but the increased firepower and more robust structure were important improvements. The Camel was not the fastest fighter and its rotary lost power at higher altitudes, but it was very maneuverable and good in a dog-fight. Unfortunately, its eccentric handling and tricky fuel system lead to more training fatalities than combat fatalities.

Below: The simple, compact cockpit of the Camel above.

Powered by the 150 hp Hispano-Suiza V-8, the SE.5 arrived at the front in April 1917 with No.56 Squadron and was a successful fighter, although it was replaced by the SE.5a with 200 hp Hispano-Suiza as soon as the more powerful engine became available. The SE.5 was thus a transitional design leading to the more famous fifth generation SE.5a.

Above Left: SE.5 A4853 "C2" of 56 Squadron, RFC flown by Lt. D.S. Wilkinson. The over-wing Lewis could be tilted to fire upward, a handy feature in combat.

Left: SE.5 A8909 "A1" of 56 Squadron, RFC flown by Capt. P.B. Prothero, leader of A Flight.

Below: SE.5 A8913 "B2" flown by Lt. Muspratt of 56 Squadron; Sgt. Vousden is posing with the fighter.

Spad 7 S.1461 of *Sous-Lt*. Rene Fonck, *Escadrille* Spa.103, May 1917. Fonck scored 75 victories to become the leading Allied ace.

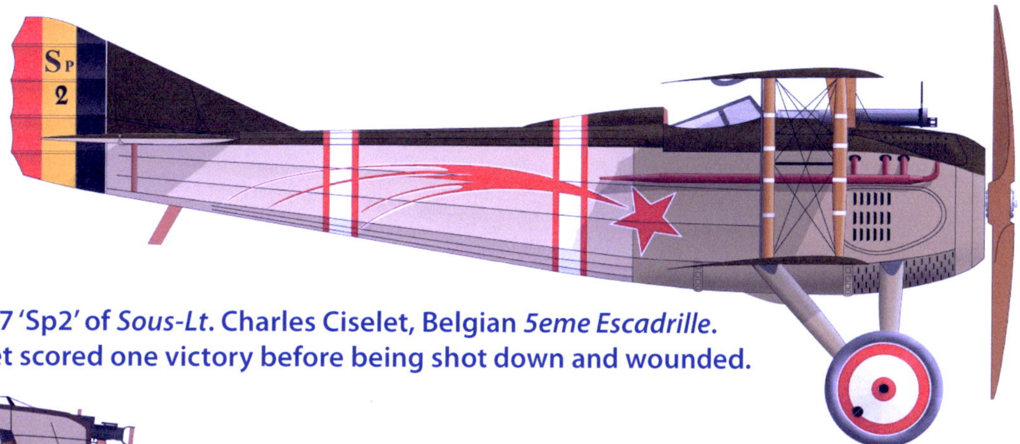

Spad 7 'Sp2' of *Sous-Lt*. Charles Ciselet, Belgian *5eme Escadrille*. Ciselet scored one victory before being shot down and wounded.

RAF SE.5 A'4850 of Capt. Albert Ball, VC, OC A Flt., No.56 Sqdn., RFC, May 1917. Ball scored 44 victories.

RAF SE.5 A'4855 of Lt. Cecil Lewis, No.56 Sqdn., RFC, April 1917. Lewis scored 8 victories and had a distinguished postwar career, co-founding the BBC and writing *Sagittarius Rising*, a memoir of his days as a WWI fighter pilot, among other notable books.

Note: SE.5 A'4855 is shown with original wind screen, which was removed before No.56 Sqdn. went into combat.

Above: With 37 victories, 35 over balloons, Coppens was the leading Belgian ace and the leading balloon-buster of the war. Coppens was said to have flown an HD.1 painted blue overall, but no photos of this aircraft have been found to confirm it.

Serving in the Belgian army when war broke out, Willy Coppens volunteered for the Belgian Air Service in September 1915 and received his pilot's certificate in December 1915. His combat flying started in April 1917 when he joined the *Sixieme Escadrille* flying BE.2c two-seaters. In July he transferred to the *1ère Escadrille de Chasse*; initially he flew the Nieuport 16, then the Nieuport 23.

When Hanriot HD.1s were offered to the squadron, he was the only pilot to accept one. He preferred the HD.1 to the Nieuport and gradually the rest of the pilots were won over to the HD.1.

Despite months of combat, Coppens did not score any victories. On March 17, 1918, he attacked his first observation balloon and forced the observers to bail out; the balloon was damaged but did not burn. On April 25 Coppens finally scored his first victory, downing a Rumpler two-seater. On May 8 he flamed two observation balloons and started his successful career as a balloon buster.

Soon after, Coppens cut a balloon's cabling loose when attacking it and it flew up and struck his Hanriot. After his plane fell off the balloon he restarted its engine and flew back to base. During a later attack Coppens landed his Hanriot on the damaged balloon and shut down the engine to protect his propeller. After the balloon started to descend he slid off it and flew away. Some time in May he received a special 11mm Vickers for his Hanriot; its larger bullets (7.7mm was standard) were more effective against balloons.

On his last mission on October 7 Coppens had downed one balloon and was attacking a second when he was badly wounded by ground fire. He crash-landed and his lower leg was amputated.

After the war he was knighted for his service. His biography, *Days on the Wing*, was published in 1931 and reissued in 1971 as *Flying in Flanders*.

During the inter-war period he was Belgian's air attaché to four different countries, and in 1928 set a parachute jump record by jumping from 19,700 feet (6,000 meters). After the German invasion of 1940, Coppens moved to Switzerland where he was active organizing resistance. In the late 1960s he returned to Belgium, where he died in 1986. Belgium's leading ace, Coppens scored all his victories in the HD.1.

Above: Ponnier Po 18 being flown at the front by the Belgian Air Service. A third-generation French design, the Ponnier was a fragile fighter and the few that were flown at the front were in Belgian service; France did not use it on operations.

Above: Belgian *Adjudant* Egide Roobaert flew this Nieuport N16 armed with 'le Prieur' rockets designed to destroy observation balloons.

Right: Belgian Hanriot HD.1 wearing the thistle marking of the *1ère Escadrille de Chasse*, Coppen's squadron.

Above & Below: The Hannover CL-types (above) and Halberstadt CL.II (below) were light C-types known as CL-types designed for escort and ground attack. With 160 hp Mercedes in the Halberstadt, or 180 hp Argus in most Hannovers, they were somewhat slower than contemporary single-seaters due to the additional weight of the gunner. However, the gunners had exceptional fields of fire and were placed close to the pilot for good coordination during combat. Moreover, the pilot had a synchronized gun and both types could maneuver with single-seat fighters, making them very dangerous opponents in air-to-air combat for any fighter during the war.

Above & Above Right: The AEG C.VII (above) and AEG C.VIII (above right) were unsuccessful CL-type competitors.

Halberstadt CL.II '5' of *Schlasta* 5, Chivres Aerodrome, August 1918. The red/white stripe was the unit marking.

Halberstadt CL.II 5720/17 '3' of *Schlasta* 21 crewed by *Uffz*. Max Neimann & *Vzfw*. Rudolf Kolodzicj, Habay La Vielle Aerodrome, October 1918. The red/yellow stripes were the unit marking.

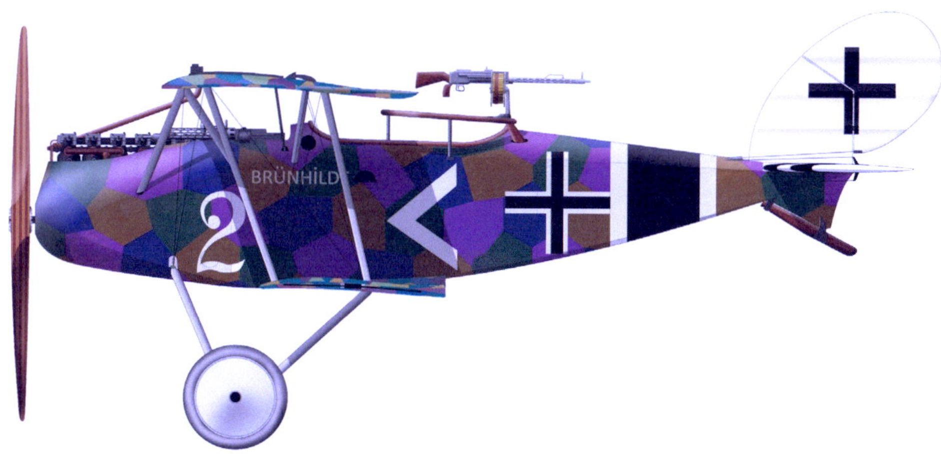

Halberstadt CL.IV '2' *Brünhilde* of *Schlasta* 27b, late 1918. The chevron and black/white/black fuselage band were the unit markings.

Hannover CL.II '4' flown by *Ltn*. Faschingbaur & *Ltn*. Stadlbauer of *Fl.Abt.(A)* 292b. The crew were KIA in this aircraft on March 6, 1918.

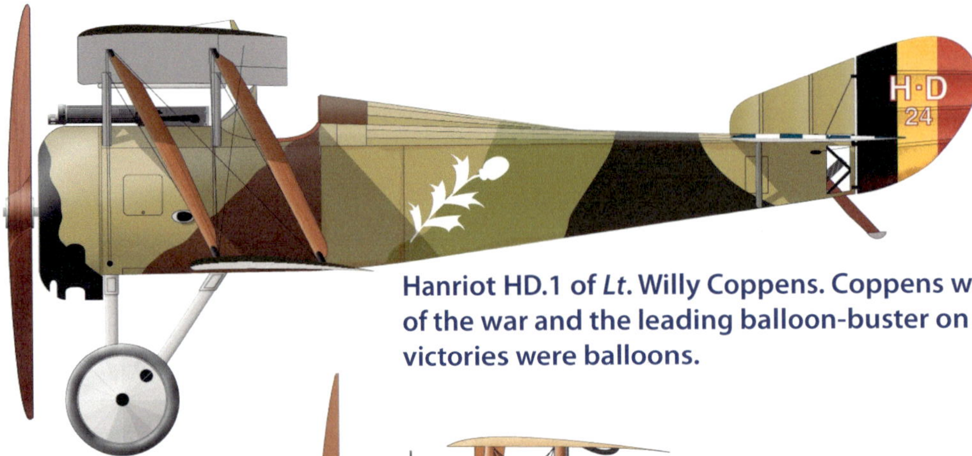

Hanriot HD.1 of *Lt.* Willy Coppens. Coppens was the leading Belgian ace of the war and the leading balloon-buster on either side; 35 of his 37 victories were balloons.

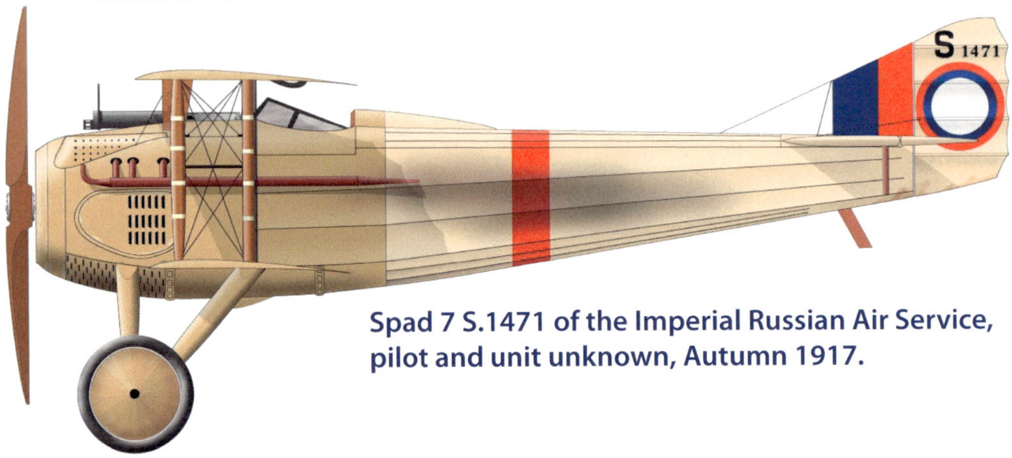

Spad 7 S.1471 of the Imperial Russian Air Service, pilot and unit unknown, Autumn 1917.

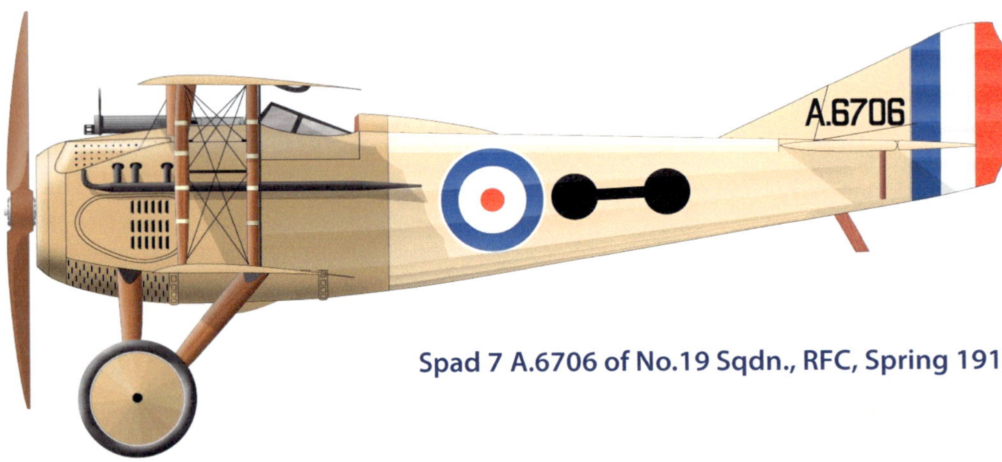

Spad 7 A.6706 of No.19 Sqdn., RFC, Spring 1917.

Camel F6034 'N' of Lt. George Vaughn, 17th Aero Squadron, USAS. Vaughn scored 13 victories flying with No.85 Sqdn., RAF and the USAS.

Above: The Hannover and Halberstadt CL-types were grouped into *Schlachtstaffels* (ground-attack units) of six aircraft and crews each. They operated individually for escort, reconnaissance, and ground-attack, and would operate as a unit as shown above during critical attacks during the land battle. Unlike many later fighter pilots, the *Schlachtstaffel* crews always remembered that the land battle was critical and their comrades on the ground depended on their help.

Below: The Halberstadt CL.IV was developed from the CL.II by shortening the fuselage to improve maneuverability; the same engine was used and performance remained essentially the same. The CL.IV arrived at the front in the early summer of 1918 and gradually replaced the CL.II in German units. The example below was restored and is now in the collection of the National Museum of the USAF.

Above: The Vickers F.B.26 Vampire appeared in May 1917. Despite its advanced 200 hp Hispano-Suiza engine, its obsolete pusher configuration prevents it from being considered a fifth-generation fighter. Several prototypes were built, this one having three Lewis guns.

Above: French *Croix de Guerre* (Cross of War) awarded to Allied soldiers who served in WWI.

Left & Above Left: The elegant Morane-Saulnier AF biplane fighter prototype was a departure from the monoplane configuration most common for Morane designs. Power was a 150 hp Gnôme rotary engine and a single synchronized gun was fitted. The AF was fast and maneuverable, but the fast, two-gun Spad 13 was already in production and only one AF was built.

The Allies used monoplanes in moderate numbers, Morane-Saulnier producing the most common types in service. All the aircraft illustrated here had thin airfoils that required extensive bracing, limiting the advantage of the monoplane configuration; contemporary biplanes were stronger for less weight and nearly the same drag. The comparative fragility of monoplanes made them unpopular and monoplane designs were generally avoided as air combat intensified.

Above: The Morane-Saulnier AI was the final Morane-Saulnier monoplane fighter. Used at the front in small numbers in early 1918, it was soon replaced by the more robust Spad and relegated to training, where it gave good service. Like other Allied monoplane designs, its thin airfoil required extensive bracing, limiting any drag advantage it might offer compared to biplanes, which were more robust.

Above: The Bristol M.1c was a fast, maneuverable fighter relegated to the Middle East and training due to lack of confidence in the strength of its monoplane configuration.

The early Roland fighters were strong, fast, and well-armed, but had mediocre flying qualities and maneuverability. The Roland D.VI was much more maneuverable than early Roland fighters and handled better, but its performance was similar.

Few big-name aces flew Rolands because they preferred – and received – fighters from other manufacturers. One notable exception was *Pour le Mérite* ace *Oblt*. Otto Kissenberth, who scored 20 victories. Kissenberth started his career as a fighter pilot in spectacular fashion, downing three aircraft during the Oberndorf bombing raid of October 12, 1916, while flying a Fokker D.II. Most of Kissenberth's victories were scored while flying Albatros fighters, but as *Staffelführer* of *Jasta* 23b he flew the Roland D.VIa for a short time before crashing on May 29 1918 while flying a captured Sopwith Camel! He was awarded the *Pour le Mérite* while in hospital and, after partial recovery, commanded the flying school at Schleissheim until the Armistice. Like many early pilots, he wore glasses as seen at right, a photo taken before his award of the *Pour le Mérite.*

Right: *Pour le Mérite* ace *Oblt*. Otto Kissenberth.

Below: Although the Roland D.VI did not win the First Fighter Competition, it made a good enough showing that 50 were ordered for operational evaluation at the front. This spectacular D.VIa was flown by *Pour le Mérite* ace *Oblt*. Otto Kissenberth with *Jasta* 23b.

Roland D.VIa of ***Lt***. Otto Kissenberth, *Jasta* 23b. Kissenberth scored 20 victories and was awarded the *Pour le Mérite*.

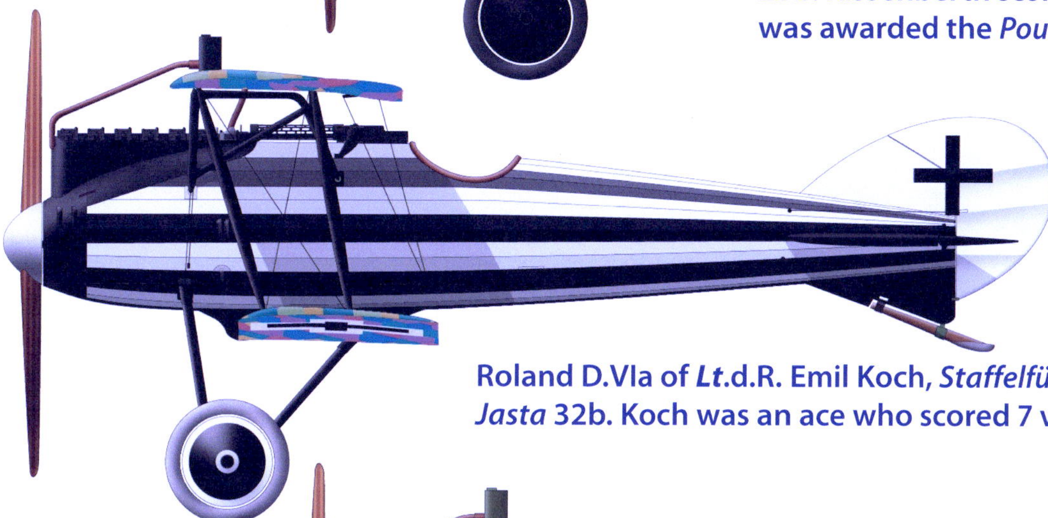

Roland D.VIa of ***Lt***.d.R. Emil Koch, *Staffelführer* of *Jasta* 32b. Koch was an ace who scored 7 victories.

Roland D.VIb 6138/18, pilot and unit unknown.

Pfalz D.XII 2525/18, *Vzfw*. Marchner, *Jasta* 35.

Above: Completed in May 1917, the AEG D.I was powered by a 160 hp Mercedes. Like all AEG aircraft it was built of steel tubing covered by fabric. Testing revealed the D.I as the fastest fighter of its day, with a speed of 225 km/h (137 mph), and its climb was equal to the new Albatros D.V that had just entered service. Its potential was obvious and *Idflieg* ordered additional prototypes for testing. On 21 August *Idflieg* test pilot *Lt.* Hendrichs was killed when his D.I crashed out of control. The cause was not determined, and flight evaluation continued on 25 August using another prototype. *Lt.* Walter Höhndorf, *Jastaführer* of *Jasta* 14, had been involved with design of the AEG D.I, and a prototype was delivered to Höhndorf at *Jasta* 14. On 5 September Höhndorf was killed flying the D.I prototype when it spun out of control. With two unexplained fatal crashes, *Idflieg* cancelled planned D.I production. The German fighter pilots needed an aircraft with the speed of the AEG D.I, which was faster than any operational German fighter of the war, so the decision to cancel it despite two fatal accidents must have been difficult. In retrospect it seems continued development was warranted due to the aircraft's potential unless there was a fundamental and irreparable flaw in the design, which does not appear to be the case. The cause of the accidents was never determined so the potential for making the aircraft safer through further development is unknown. However, the small number of production aircraft were flown for some time, including as comparison aircraft in the First Fighter Competition, without further fatalities, bringing its cancellation into question.

Right: The Kondor D.VI, powered by a 110 hp Oberursel U.II, was an attempt to improve the pilot's field of view forward and upward. Eliminating the upper wing center section undoubtedly resulted in more induced drag than a conventional wing, which limited its speed and climb, and would have significantly reduced its performance during high-G maneuvering combat. It remained a prototype.

Pfalz D.XII in the Australian War Memorial after restoration (courtesy Colin Owers)

Above & Below: The Pfalz D.XII followed the Fokker D.VII into service, but is harder to categorize. It used the same structural technology and engine as the Pfalz D.IIIa, other than the wing cellule. So, does this make it a third- or fourth-generation fighter? Some D.XII fighters were allocated the advanced BMW engine at the factory, granting it fourth-generation status, but no photos have been found showing an operational D.XII with a BMW; perhaps all the BMW engines were cannibalized for the superior Fokker D.VII? So the D.XII, which served in significant numbers in late 1918, was one of those aircraft, like the Sopwith Camel, half-way between the generations. Below, *Vzfw*. Marchner of *Jasta* 35.

Germany's Triplane 'Craze'

Above: The cause of all the trouble, the Sopwith Triplane. Clearly related to the preceding Pup, the Triplane had an elegant triplane wing cellule of high aspect ratio with enough gap between the wings to minimize airflow interference between them. The single fixed gun remained, but the engine was initially a 110 hp Clerget 9Z, with most production aircraft powered by a 130 hp Clerget 9B, an increase of 62.5% over the power enjoyed by the Pup. The Triplane had clean lines, good maneuverability, and good handling qualities, but was not as robust as the later Sopwith Camel biplane. The middle wings were attached to the cabane struts instead of directly to the fuselage, creating additional drag from vortices at the wing roots as well as the wing tips. Ailerons were fitted to all wings for a good roll rate.

Introduction of the powerful Albatros fighters armed with two synchronized machine guns in the late summer of 1916, coupled with the formation of the first German fighter squadrons, gave the German Air Service superiority over the Western Front. Organization of fighter units matched the moves pioneered by the French early in 1916 over Verdun and soon adopted by the RFC and RNAS, while the Albatros returned the technical advantage to Germany, who had lost it when the French Nieuport fighters and the British DH-2 appeared over the front to challenge the Fokker and Pfalz monoplane fighters.

As more Albatros fighters were delivered and more Jastas were formed, German air superiority gained momentum, peaking in April 1917 with a sound tactical defeat of the Allied air arms in a debacle named "Bloody April" by the hard-pressed British flying services.

During this period of German resurgence over the Western Front, two new Allied fighters were introduced in small numbers that presaged Allied recovery. One was the Spad 7 biplane. Powered by the innovative new Hispano-Suiza V-8, the Spad was strong, fast, and maneuverable. Its main limitation as a fighter was that it only carried one synchronized machine gun. Building up to a significant production rate was also a problem, with the new engine being the main problem.

While the French introduced the sturdy Spad biplane, the British RNAS introduced a more distinctive fighter. For sound structural and aerodynamic reasons, most airplanes before and during the war were either monoplanes or biplanes. However, in early 1917 Sopwith introduced a triplane development of their agile Pup fighter. The Sopwith Triplane retained the single gun and general lines of the Pup, but had a more powerful, 110 hp

Above: This flying reproduction Fokker Triplane is in authentic colors for Manfred von Richthofen, the famous Red Baron. Prototype Fokker Triplanes arrived at the front in September 1917 and the Red Baron was killed in his triplane on 21 April 1918. This connection with the famous Red Baron makes the Fokker Triplane far more famous than it would be otherwise, and arguably makes it the most famous airplane of the Great War.

rotary coupled with a new triplane wing cellule.

Apart from their single gun armament, both the Spad and Triplane out-performed the Albatros. However, neither was available in sufficient numbers to prevent the Albatros from achieving air superiority in late 1916 and early 1917. While both the Sopwith and Spad designs were good, the Spad 7 had more development potential and its two-gun development, the Spad 13, remained at the front until the armistice.

Apparently it was the Sopwith's unusual triplane configuration that was the key to its charm. While a good fighter, the Spad was a biplane, the most common configuration in WWI, and its performance could be attributed to its innovative engine. On the other hand, the Sopwith Triplane was the first triplane fighter to appear over the front, and its rotary engine was a known quantity. Regardless, the Sopwith Triplane caught the imagination of Idflieg, who imagined that its triplane configuration was the secret to its success.

Despite the overall German superiority during the Spring of 1917, the Sopwith Triplane caused Idflieg to think that perhaps the triplane had some inherent superiority to the biplane. To the contrary, the biplane configuration is inherently a better compromise between strength, weight, and aerodynamic drag than the triplane, which has too much drag.

However, the light, agile Sopwith Triplane had hit a lucky peak of triplane potential, and the result was the triplane craze, an affliction confined almost entirely to Germany and Austria. German authorities, followed by the Austrians, requested their manufacturers to develop triplane fighters, resulting in a plethora of new designs that never reached the Front and wasted a great deal of engineering resources in the process. Only two German triplane fighters reached production. One was the Pfalz Dr.I (Dr for *Dreidecker*, or triplane);

only ten were built and while there are photographs of it at *Jasta* 73, its appearance in combat is not confirmed. The other was the Fokker Triplane, made famous as the airplane the Red Baron was flying when he was killed.

In many ways the Fokker Triplane was a paradox. It came at a time when the Albatros, Roland, and Pfalz fighters were inferior to their Allied opponents, so was welcomed warmly by the German pilots. Initially beset with some structural failures due to poor quality control during production, the problems were soon rectified and it became a favorite mount of German aces despite being slow, its main limitation, due to its low powered rotary engine.

On the positive side, the Fokker Triplane introduced Fokker's cantilever wing design to become Germany's first fourth-generation fighter. Its innovative wing gave it exceptional climb and maneuverability ideally suited to close-in, high-G maneuvering combat, or dog-fighting as it was called, that was so characteristic of WWI combat over the Western Front. It had the standard two synchronized machine guns and was, once the quality control problems were solved, a sturdy airplane. Despite the claims of some critics, it was not a copy of the Sopwith, sharing nothing in common with the Sopwith's design other than its triplane configuration and use of a rotary engine. While it needed more speed to be a first-class fighter, at least it out-performed its opponents in climb and maneuverability, whereas the Albatros was now inferior in all respects.

Above: Only one Astoux-Vedrines triplane was built. This French fighter was powered by a 130 hp Clerget rotary.

Eventually sanity prevailed, and designers abandoned the triplane configuration due to its higher drag. The Sopwith Triplane was replaced by the biplane Camel, which did not really out-perform it but was much sturdier and had two guns. The Fokker Triplane was developed into the faster Fokker D.VI biplane. But the D.VII, powered by a water-cooled Mercedes or BMW, was a more potent fighter with higher ceiling and better speed and climb at altitude, and D.VI production was limited.

Above & Left: Fokker Triplanes of *JG*II. The Fokker Triplane is difficult to categorize; it had a low-power engine and retrograde triplane configuration, but was the first operational aircraft to feature Fokker's advanced wing design, which makes it a fourth generation fighter despite its other attributes.

Sopwith Camel B6313 of Capt. William Barker, OC 'C' Flt., No.28 Sqdn., RFC, Grossa Aerodrome, Italy, late 1917. Barker scored 50 victories and was awarded the Victoria Cross.

Sopwith Camel 'Sc7' of *Lt.* Jean Olieslagers, *9me Escadrille*, Belgian Air Service. Olieslagers scored 6 victories.

Nieuport 28 N6157 *Becky* of 1/Lt. William J. Hoover, 27th Aero Sqdn., USAS, July 1918. Hoover scored 3 victories and was awarded the DSC.

Austin Osprey X15, Prototype February 1918; Britain did not abandon triplane fighters until well into 1918.

Fokker Dr.I 214/17 flown by *Ltn.d.R.* Wilhelm Papenmeyer, *Jasta* Boelcke, March 1918. Papenmeyer scored 4 victories before he was KIA. At this time the *Jasta* Boelcke unit colors included the tailplane painted white on one side and black on the other.

Fokker Dr.I 152/17 was one of the Fokker Triplanes flown by Manfred von Richthofen, commanding *JGI*. It is not as well-known as the all-red Triplane in which he was killed.

Fokker Dr.I 586/17 as flown by *Ltn.* Ernst Udet, who added his fiancee's initials to the side of the aircraft. Udet scored 62 victories and was awarded the *Pour le Mérite*. Udet became a famous movie and airshow pilot after the war.

Fokker Dr.I 419/17 flown by *Ltn.* Walter Göttsch, *Staffelführer* of *Jasta* 19. Göttsch scored 20 victories and was KIA on 10 April 1918.

Above: Pfalz D.III triplane conversion powered by a 160 hp Mercedes D.III engine on the Speyer airfield in October 1917. Ailerons were fitted to both top and bottom wings to improve the roll rate.

Above: Pfalz Dr.I 3050/17, the second prototype, showing its armament. The Dr.I aircraft all were finished in the then-standard Pfalz silver-gray with standard iron crosses without white borders.

Above: Reproduction Fokker Triplane in authentic camouflage and markings for Lothar von Richthofen of *Jasta* 11. Lothar, a 40-victory ace, was the Red Baron's younger brother. The red struts, wheel covers, and cowling are the *Jasta* 11 markings; the yellow tail and top wing were Lothar's personal markings.

Left: Manfred (at right) and Lothar (at left) von Richthofen in March 1918. Both are wearing the *Pour le Mérite*, also known as the Blue Max, around their necks.

Facing Page, Top: During the German triplane craze, no biplane fighter was safe from modification. Even the Albatros D.V was fitted with a triplane wing cellule for comparison, creating the Albatros Dr.I. The standard biplane D.V was faster, had better climb, and was more maneuverable due to the extra weight and drag of the triplane wings. (Both photos The Peter M. Bowers Collection/The Museum of Flight)

Facing Page, Bottom: *Hptm*. Wilhelm Reinhard by his Fokker Triplane while serving with *Jasta* 11. He became commander of the Flying Circus after Richthofen was killed. While testing the Zeppelin-Lindau D.I all-metal fighter on 3 July 1918, Reinhard was killed when the upper wing broke off. With 20 victories he had been nominated for the *Pour le Mérite*, but his accidental death prevented the award. His accident inspired part of the plot of *The Blue Max*, a novel and movie about WWI German fighter pilots by Jack D. Hunter.

Above: The Naglo D.II fighter prototype, either serial 1161/18 or 1165/18, was evaluated in conjunction with the Second Fighter Competition. The bottom wing was mounted on a keel built into the fuselage.

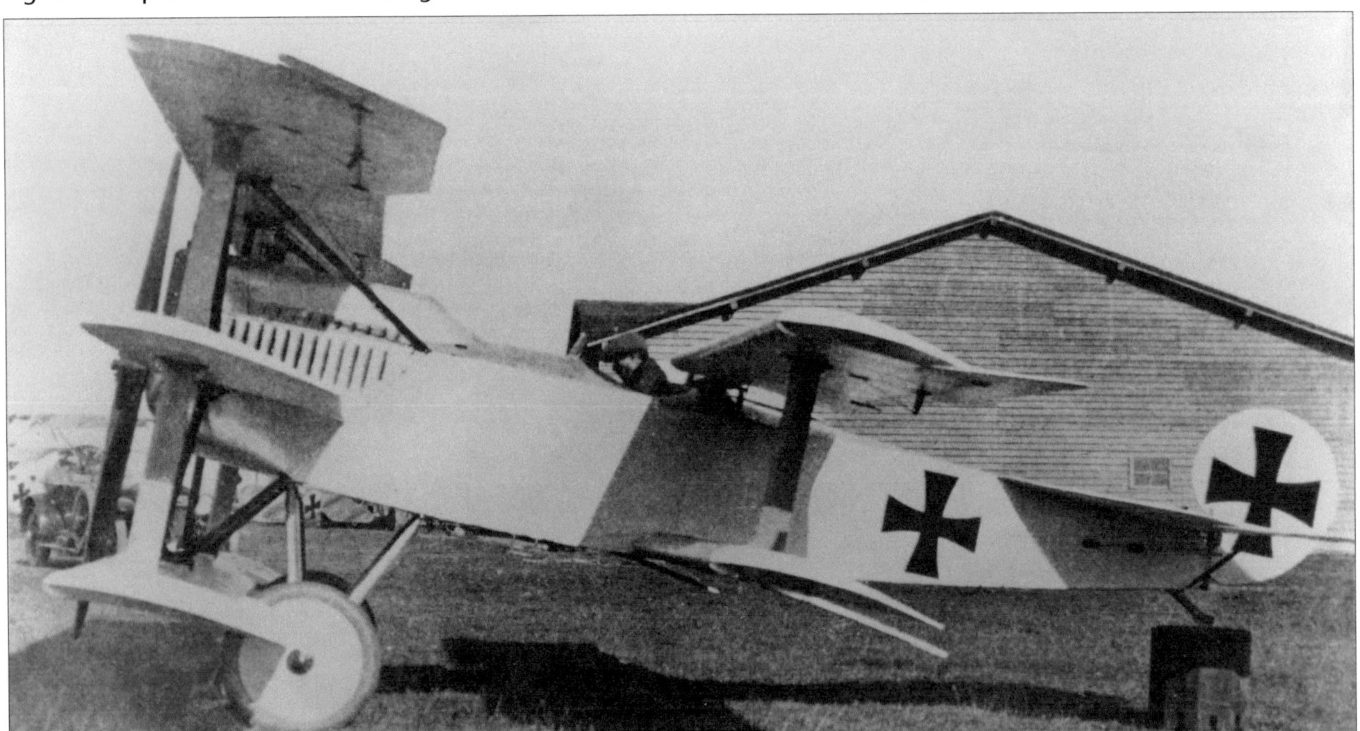

Above: The bizarre Fokker V8 prototype used a Triplane wing cellule in front and the upper two wings of a Triplane aft. Two short test flights proved the V8 had dangerous handling qualities and was immediately abandoned. WWI aircraft development was mostly empirical, not theoretical, and the fact Fokker tried this five-wing monstrosity shows that the secret to his success was a constant series of experiments — many of which did not work! Proof that even the best designers can get carried away, the Fokker V8 exceeded the limits of good judgment. The old aviation axiom that, 'if it looks right it will fly right', is usually, although not always, true. However, there is a corrollary; 'if it looks wrong, it will not fly right', and the Fokker V8 is a prime example. Learning his lesson, Fokker recovered from this debacle and went on to design the exceptional D.VII biplane and innovative E.V/D.VIII monoplane fighters.

Above: The SSW DDr.I used two 110 hp Sh.I engines, one mounted in front of the nacelle and the other mounted in the rear of the nacelle. Two synchronized guns were fitted. The eccentric design crashed on its first test flight before performance data could be recorded.

Above: Austria was affected too; this view of Aviatik 30.24 during flight testing at Aspern in October 1917 clearly shows it used an Aviatik D.I fuselage and tailplane modified to take a new triplane wing cellule. As was generally the case with triplane conversions of biplane designs, the additional weight and drag of the third wing reduced speed and climb compared to the standard D.I biplane.

Above Right: The idea that more wings could improve performance reached its height with the quadraplane. The Friedrichshafen FF54 is shown here in its original quadraplane form. The narrow gap between the pair of middle wings created airflow interference between them and contributed to the aircraft's failure. The lower middle wing was then removed and the vertical tail redesigned for renewed flight testing as a triplane. However, the FF54 was not successful as a triplane either.

Above: Triplanes were even tried for floatplane fighters; SF4 Marine #901 was developed from the SF4 #900 biplane.

Fifth Generation Fighters

By our definition, a new fighter must have innovative new engines or structural and aerodynamic designs to qualify as a fifth generation WWI fighter.

For Allied fighters, the innovations were primarily powerful, water-cooled V-8 or V-12 engines like the Hispano-Suiza V-8 or Rolls-Royce V-12. The Bentley B.R.2 rotary, the most powerful rotary engine to serve in the war, similarly qualified the Sopwith Snipe as a fifth-generation fighter.

For the German fighters, the structural and aerodynamic advances came first with Fokker's innovative cantilever wings, followed by the Sh.III counter-rotary engine and the BMW.IIIa, then the Mercedes D.IIIaü over-compressed engines.

The SE.5a, powered by a 200 hp Hispano-Suiza engine, was derived from the 150 hp SE.5 and was one of the most successful fifth-generation fighters. Similarly, the Spad 13, initially powered by a 200 hp Hispano-Suiza, was derived from the earlier Spad 7 and became the standard front-line fighter in French and American service and was also used by Belgium, Great Britain, and Italy in small numbers. The Spad 13 and SE.5a were the most numerous and important Allied fifth-generation fighters.

Fifth Generation Fighters	At Front
Bristol F.2A/B	3/17– Armistice
Fokker D.VI	6/18–Armistice
Fokker D.VII	5/18–Armistice
Fokker E.V/D.VIII	8/18–Armistice
Pfalz Dr.I	4/18–8/18
Pfalz D.VIII	5/18–Armistice
Pfalz D.XV	To *Flugparks* week of Armistice
SE.5a	6/17–Armistice
Sopwith Dolphin	2/18–Armistice
Sopwith Snipe	10/18–Armistice
Spad 12	8/17–Armistice
Spad 13	4/17– Armistice
Spad 17	~7/18– Armistice
SSW D.III	4/18–Armistice
SSW D.IV	8/18–Armistice

The Fokker D.VII, with its thick wood cantilever wing, was by far the most important German fifth generation fighter. By the summer of 1918 it was the most numerous German fighter at the front and desired by all German fighter pilots. The excellent SSW fighters, which exhibited unsurpassed climb and maneuverability, served in only small numbers.

Above: without doubt the SE.5a was one of the most significant and successful fifth-generation fighters. It served in large numbers and was the best British single-seat fighter in combat. Not a dog-fighter like the Camel, the SE.5a was fast, robust, handled well, and was a safer aircraft. Here James McCudden, one of the leading British aces, sits in his personal SE.5a in No.56 Squadron. A former mechanic, McCudden constantly tinkered with his SE.5a to improve its performance, which enabled him to intercept the high-flying Rumpler reconnaissance airplanes that the standard SE.5a and other Allied fighters could seldom reach.

Right: The De Bruyère pusher fighter was another French attempt to create a pusher fighter of low drag and high speed without the need for synchronizing gear. Armed with a 37mm cannon, it was tested in April 1917, well after France had fielded the Spad 7 fighter with synchronized gun. Power was from a Hispano-Suiza V-8 of unknown type and power. The pivoting wingtips may have caused the crash of the prototype on its first flight in April 1917.

Above: The distinctive Bristol F.2B two-seat fighter was powered by a 275 hp Rolls-Royce V-12. Originally designed as a reconnaissance aircraft, it had good maneuverability and its powerful V-12 gave it good speed and climb; crews quickly learned to fly it offensively like a fighter in combat instead of defensively like a typical reconnaissance plane. Flown offensively, it was one of the greatest aircraft of the war and produced many aces, some of them gunners. Production of the complex V-12 depended on excellent craftsmanship, limiting production to far less than the demand for this excellent engine. Eventually, less powerful 200 hp V-8 engines were installed in the airframe for use as reconnaissance planes while the Rolls-Royce-equipped aircraft were used for offensive patrols. Even with the lower-power engines the Bristol was superior to other British reconnaissance aircraft of its time.

Above: Werner Voss by his Fokker F.I 103/17, one of the first two Triplanes at the front, the other going to Manfred von Richthofen. This is the aircraft he was flying during the famous dogfight on 23 September 1917. Voss had 48 victories at the time of his death and was 20 years old.

Above: Arthur Rhys Davids, credited with downing Voss on 23 September 1917 for his 21st of an eventual 27 victories. Rhys Davids, who lamented not being able to bring Voss down alive, was himself killed in action on 27 October by *Lt.* Karl Gallwitz of *Jasta* 2 for his 5th victory and the first of three British aces he claimed in his total of 10 victories. Rhys Davids was 20 years old.

The most celebrated dog-fight of the war occurred on 23 September 1917 between 48-victory ace *Lt.* Werner Voss and two flights of aces from No.56 Squadron. Voss, who was only four victories behind his friend and competitor Manfred von Richthofen, was making one last flight before going on leave. Voss first attacked and damaged two SE.5as of No.60 Squadron who were forced to dive away to escape.

No.56 Squadron then arrived to the rescue, with aces James McCudden, Authur Rhys Davids, and others flying SE.5a fighters, and attacked Voss. The results were surprising; Voss out-maneuvered all six SE.5as and fought them for more than eight minutes.

McCudden commented that Voss turned 'in a most disconcertingly quick manner, not a climbing or Immelmann turn, but a sort of flat half spin'. The triplane was now in the middle of the SEs, but none could get in an effective burst of fire: 'its handling was wonderful to behold. The pilot seemed to be firing at us all simultaneously, and although I got behind him a second time, I could hardly stay there for a second. His movements were so quick and uncertain that none of us could hold him in sight at all for any decisive time.'

Bowman later recalled the triplane pilot was now alone, fighting six SEs, 'which did not appear to deter him in the slightest. At that altitude he had a much better rate of climb, or rather zoom, than we had and frequently he was the highest machine of the seven and could have turned east and got away had he wished to, but he was not that type and always came down on us again. His machine was exceptionally manoeuvrable and he appeared to be able to take flying liberties with impunity.'

Keith Muspratt was the next to be put out of the fight: his machine was hit and he was forced to return.

Above: Robert Chidlaw-Roberts was the second of two 60 Squdron SE.5a pilots forced down by Voss before 56 Squadron arrived. Credited with 10 victories, Chidlaw-Roberts, seeing Hamersley, the first 60 Squadron pilot forced down by Voss, was in trouble, dived to attack Voss in an attempt to draw him away from Hamersley, but the enemy pilot hardly paused in his pursuit: 'In seconds he was on my tail and had shot my rudder bar about. I retired from the fray and that is all I saw of it.'

Above: Formal portrait of Werner Voss.

Bowman reported: 'I, myself, had only one crack at him: he was about to pass broadside on across my bows and slightly lower. I put my nose down to give him a burst and opened fire, perhaps too soon; to my amazement he kicked on full rudder, without bank, pulled his nose up slightly, gave me a burst while he was skidding sideways, then kicked on opposite rudder before the result of this amazing stunt appeared to have any effect on the manoeuvrability of his machine.'

Distracted by Bowman, Voss was now flying straight for a brief moment and Rhys Davids was behind him with, as Bowman observed, 'his prop boss almost on the tail of the triplane, flying nose down and straight for the first time in the entire fight.'

At dinner that night in the 56 Squadron Mess, the fight with the enemy triplane was the entire topic of conversation. Rhys Davids was showered with congratulations, but his only reply was: 'Oh, if only I could have brought him down alive.' All the pilots who had taken part in the fight shared his regret. Bowman recalled: 'Our elation was not nearly as great as you might have imagined. Rhys Davids, I think, was genuinely upset.' There was a great deal of speculation as to who might have been flying the enemy triplane. It had obviously been one of the very best German pilots and opinions differed as to whether it had been Wolff, Voss, or even von Richthofen. Speculation was also rife in the 60 Squadron Mess at Ste-Marie-Cappel, but from precious experience 'Grid' Caldwell had little doubt that it had been Werner Voss. 'I, for one, felt it could be Voss as I had experienced him before. He was a terrific chap, and rated easily No.1, including von Richthofen.'

RAF SE5a D213 of 2/Lt P.D. Learoyd, No.40 Sqdn., RFC, Spring 1918

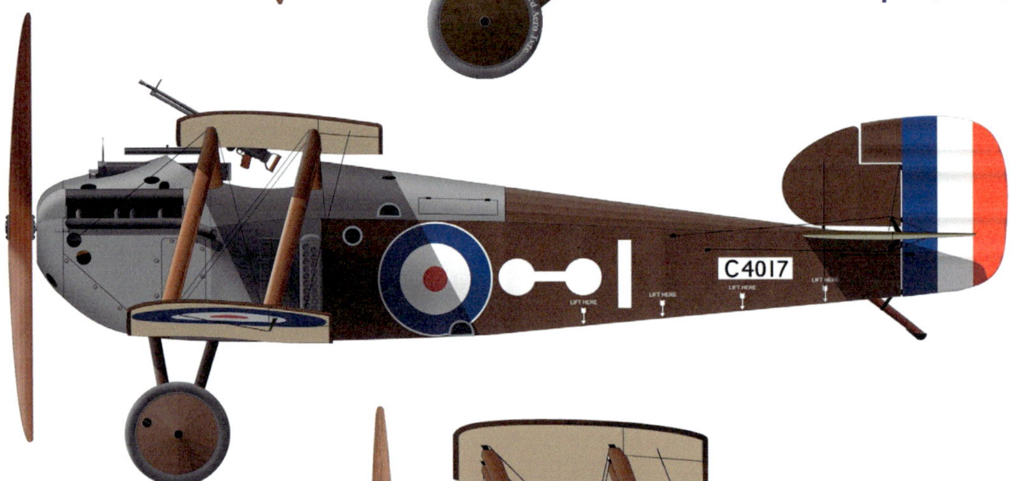

Sopwith Dolphin C4017 flown by Major A.D. Carter, No.19 Sqdn., RAF, Savry Aerodrome, April 1918. Carter scored 29 victories, 11 of them in this Dolphin.

Sopwith Snipe E8102 flown by Major William Barker during his Victoria Cross action on 27 October 1918. Barker scored 50 victories.

Spad 12 of *Capt*. Georges Felix Madon, *Escadrille* Spa.38. Madon scored 41 victories.

Above: The Spad 12 was derived from the Spad 7 to increase firepower; the single synchronized gun fitted to the Spad 7 was its biggest weakness. The Spad 12 retained the synchronized gun of the Spad 7 and added a single-shot 37mm cannon mounted in the 'V' of the cylinders and firing through the hollow propeller shaft. The pilot could hand-load additional 37mm shells in flight. One 37mm hit could down an aircraft, but that was very difficult with a single-shot weapon and only the most skilled pilots accomplished it. Few Spad 12 fighters were used in combat, but this one was used by Nungesser at Spa.65.

Below: The Spad 13 with two synchronized machine guns was the next logical progression of the basic Spad 7 configuration. The additional firepower was important, especially if a gun jammed in combat, and a more powerful Hispano-Suiza engine was fitted to compensate for the weight of the additional gun. The Spad 13 was used by most Allied air services and became the principle fighter of the French and American air services.

Above: The Spad 17 was developed from the Spad 13 and the 20 production aircraft built used pilot production versions of the new 300 hp Hispano-Suiza engine. The production Spad 17s were assigned to the elite GC 12, The Storks. The additional power gave little more performance as the earlier, less powerful Spad 13, indicating that the airframe had reached the limit of its development potential.

Below: The Hispano-Suiza engine was also used in the British SE.5a, one of the iconic fighters of the war, and also in the distinctive Sopwith Dolphin as shown below. The Dolphin was a departure from Sopwith's series of rotary-engined fighters like the Pup, Triplane, and Camel; it was a larger aircraft intended for better performance at high altitude. The wings were staggered backward and arranged to give the pilot the best possible field of view forward and upward, and it was designed to carry two synchronized guns and one or two Lewis guns firing upward. To minimize weight normally only a single Lewis was mounted as shown on this Dolphin, and many Dolphins omitted the Lewis entirely in favor of the two synchronized Vickers. The Dolphin may have had the best high-altitude performance of any operational British fighter, but it was not very popular with pilots, who generally disliked its configuration. Fearing fatal injuries in case of over-turning on landing, most Dolphins were fitted with roll-over hoops on the upper wings as shown.

Spad 12 of *Capt.* Georges Guynemer, *Escadrille* Spa.3, July–August 1917. Guynemer scored four victories in this fighter.

Spad 13 of *Capt.* Georges Guynemer, *Escadrille* Spa.3, September 1917. Guynemer scored 53 victories before he was KIA on 11 September 1917.

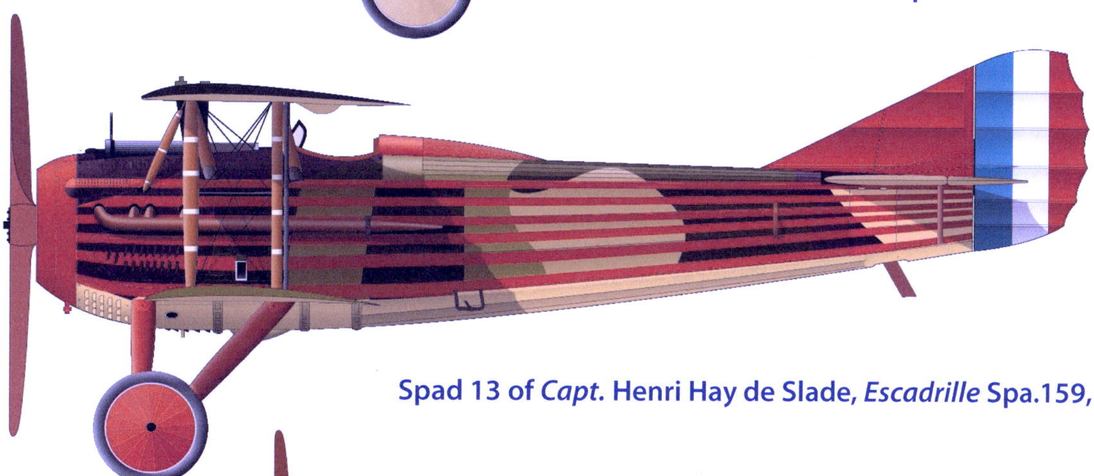

Spad 13 of *Capt.* Henri Hay de Slade, *Escadrille* Spa.159, July 1918.

Spad 13 S.15155 of 1/Lt. Frank Luke Jr., 27th Aero Squadron, USAS. Luke scored 18 victories, including 14 balloons, making him the leading American balloon-buster, before he was KIA on 29 Sept. 1918. Luke was one of two USAS fighter pilots awarded the Medal of Honor.

Bristol F.2A A3335 '4' of No.48 Sqdn., RFC.

Bristol F.2B A7288 '7' of Lt. Andrew McKeever, No.11 Sqdn., RFC, Autumn 1917. With 31 victories, McKeever, a Canadian, was the leading two-seat fighter ace of the war.

Sopwith Dolphin C4131 'T' of No.79 Sqdn., RAF, 1918.

SE.5a B'4891 of Capt. James McCudden, No.56 Sqdn., RFC, late 1917. McCudden scored 57 victories and was awarded the Victoria Cross. He was killed in a take-off accident en route to command No.60 Sqdn. He added the red spinner from an LVG C.V he had shot down.

Above: A Sopwith Dolphin without an upward-firing Lewis and without the roll-over hoops on the upper wings.

Below: Arriving in time for four or five weeks of combat, the Snipe benefitted from having the most powerful rotary engine to see combat during the war, the 230 hp Bentley B.R.2, which gave it excellent climb and ceiling. However, the Snipe experienced prolonged development. The Snipe was an enlarged, more powerful derivative of the Sopwith Camel, and that was its limitation. Early Snipe prototypes had single-bay wings, but the greater engine weight forced the Snipe to use larger, two-bay wings for sufficient strength. In turn, the heavier two-bay wing created more drag from its more numerous struts and bracing wires, with the result the Snipe was little or no faster than the Camel, the aircraft it was to replace, despite its greater power. The Snipe was also slower than the Sopwith Dolphin already in service, although it could out-climb both the Camel and Dolphin due to its powerful rotary. The Snipe was not as agile as the Camel, and the Snipe prototypes underwent an extensive series of tests with different flying surfaces to improve maneuverability. With the arrival of the Fokker D.VII at the front, Snipe development continued in a futile attempt to match the Fokker's flying qualities. This was doomed to failure due to the thin airfoil sections used; their sharp leading edges resulted in abrupt stalling characteristics and marginal behavior at high angles of attack. Essentially, the Snipe used the structural and aerodynamic technology of 1916 coupled with the most powerful rotary engine of the war. The Fokker D.VII used breakthrough structural and aerodynamic technology for its wings, which featured a relatively thick airfoil with excellent structural and stalling characteristics, and no fighter using a thin airfoil was going to match it. Of course, the Fokker also needed a better engine than the Mercedes D.IIIa that powered the early examples, and soon the 185 hp, over-compressed BMW.IIIa provided that additional power, especially at high altitude where it was most needed.

With 75 confirmed victories, René Fonck was both the leading French ace and the leading Allied ace. He was an exceptional marksman and twice scored an amazing six victories in one day. He was also a cautious air fighter; claiming that only a single enemy bullet ever hit one of his aircraft.

However, despite his unquestioned ability and great success, he was never popular with his fellow pilots nor the French public. Fonck was reclusive and ascetic; his approach to combat was clinical and he stayed fit through calisthenics. Unable to connect with others, he constantly discussed his accomplishments, once saying, "I put my bullets into the target as if by hand." Even his best friend, Lt. Haegelen, (a 22 victory ace), was quoted as saying of Fonck: "He is not a truthful man. He is a tiresome braggart, and even a bore, but in the air, a slashing rapier, a steel blade tempered with unblemished courage and priceless skill... But afterwards he can't forget how he rescued you, nor let you forget it. He can almost make you wish he hadn't helped you in the first place."

After the war he sought to be the first to fly the Atlantic. In September 1926 his tri-motor S.35 crashed and burned on take-off, killing two of the three other crewmen. Before Fonck could try again, Lindberg succeeded. Fonck returned to military aviation and from 1937 to 1939 was Inspector of French fighter forces.

Above: Formal portrait of René Fonck.

Below: René Fonck in front of his Spad 13 of Spa.103.

Above: James McCudden, VC, in front of the experimental Vickers F.B.16D fighter he used as a personal transport while instructing in the UK in 1918, before returning to France to Command No.60 Squadron and his fatal crash.

Modest ace James McCudden, VC, is a refreshing antidote to Fonck. McCudden rose from the ranks, starting as a mechanic, then flying volunteer missions as an observer and gunner, then going to flight school to become a pilot, then becoming a fighter pilot. He was aggressive yet methodical, and constantly worked on his aircraft alongside his mechanics to get the best performance and reliability his aircraft could deliver.

McCudden was a likable young man of good character and was popular with his peers and superior officers. Their admiration grew as McCudden gained more skill and experience as his victories grew. He became an ace flying DH.2 pushers with No.29 Squadron, then returned to England for a tour as a flight instructor.

McCudden returned to France with the new No.56 Squadron flying SE.5, then SE.5a fighters. He hit is stride with this high-performance fighter and started scoring regularly. Realizing the importance of long-range reconnaissance, he started specializing in stalking and downing these aircraft. This strategic insight made his victories especially valuable. In December 1917 he alone scored most of the victories of No.56 Squadron, and he scored steadily until he returned to England for another tour as a flight instructor in March 1918. By then he had been commissioned an officer and awarded the Victoria Cross. He was well-respected as a flight instructor and flew a wide variety of aircraft. During this time he was loaned a prototype fighter, the Vickers F.B.16D, as a personal aircraft to fly for visits to various airfields as he instructed new pilots on fighter tactics.

With 57 confirmed victories (of 88 claims), McCudden officially ranks as the fourth highest-scoring British ace behind Bishop, Collishaw, and Mannock. In reality, he was probably the leading British ace in that most of his victories could be confirmed by the wreckage of an enemy aircraft, something that cannot be said for the others. Only 10 of his 88 claims were for aircraft "OOC", or 'out of control', while 'OOC' claims figured prominently in the scores of the others. British authorities were liberal in confirming OOC claims to encourage pilots to fight offensively over German lines, but the vast majority of aircraft claimed as 'OOC' had disengaged from the fight and lived to fight another day. McCudden almost certainly shot down more aircraft than any other British ace.

In July 1918 he returned to France to Command No.60 Squadron. Taking off from an intermediate stop, his engine failed and his SE.5a stalled and crashed. He had developed the ultimately fatal habit of a steep climbing turn immediately after lifting off; when his engine failed, he was too low to recover from the resulting stall. This was a tragic end for an otherwise exceptional airman and fine man.

Above & Below: The Austin-Ball A.F.B.1 fighter was the brain-child of British ace Albert Ball, who designed it in April 1916. Ball's favorite method of attack was to fly underneath an enemy aircraft and fire upward into it using the Lewis gun on his Nieuport (or later, on his SE.5) that could be raised and lowered. This influenced the A.F.B.1's armament; one Lewis was mounted above the wing in a Foster mounting fix to fire above the horizontal, and another Lewis was mounted in the vee of the 200 hp Hispano-Suiza and fired through the propeller hub without need for a synchronizing gear. The upper wing was mounted at the pilot's eye level for excellent field of view upward. Unfortunately, the A.F.B.1 was not ready for testing until July 1917, two month's after Ball's death. When tested the A.F.B.1 demonstrated exceptional speed and generally good handling qualities. However, by that time the Sopwith Camel and SE.5a were in mass production and squadron service. The Air Board determined that its performance advantage over the SE.5a, which used the same engine, was not sufficient to warrant disrupting production of the SE.5a to start production of the A.F.B.1 and no production of this excellent fighter was undertaken.

Right: The Laboratory Eiffel LE fighter was an experimental aircraft intended to test advanced aerodynamic features. The wing spars were made of duralumin and were nearly cantilever; however, bracing struts were needed. Powered by a 180 hp HIspano-Suiza, it crashed fatally on its second flight. Although the potential performance was estimated to be very high, this fatal crash ended development. Despite the crash being officially attributed to pilot error, the authorities seem to have reservations about its monoplane configuration, and its primary impact seems to have been the postwar abandonment of the low-wing cantilever monoplane by France for many years.

Right: Once Nieuport realized their traditional sesquiplane configuration had reached the limit of its potential, the search for a new configuration began. This elegant monoplane was powered by a 150 hp Gnôme rotary and was armed with two synchronized guns. The windows in the wing roots to provide a downward view for the pilot are clearly shown. No production was undertaken but development continued with the Nieuport 31. The Nieuport monoplanes were probably fourth generation of fighters.

Right: The Nieuport 29 biplane powered by a 300 hp Hispano-Suiza was the next Nieuport fighter produced in quantity, but arrived too late for combat.

Above & Above Right: Two portraits of Manfred von Richthofen, the famous Red Baron, taken at the same sitting. Now a legend, he was the war's leading ace and certainly one of its best fighter leaders. He trained his men as well as lead them, and drummed into them the lessons he had learned from Oswald Boelcke, his teacher. He also served his country by suggesting the fighter competitions that were instrumental in getting the superior Fokker D.VII to the front.

Manfred von Richthofen, known as the Red Baron, is the only soldier of the Great War generally known today, if only as a legend. Who was this archetypal warrior and fighter ace?

Manfred was born to a family of the minor nobility (hence the 'baron') in Breslau, Prussia in May 1892; he was the second child but oldest boy in the family. His younger brother Lothar was born in 1894. Manfred enjoyed his childhood, excelling in swimming, gymnastics, and other athletic pursuits. His greatest love, however, was hunting, a skill he learned early. At 11 his father enrolled him in the Cadet Acadamey at Wahlstatt, something Manfred was not enthusiastic about. In November 1912 he was commissioned an officer in the cavalry, a branch he chose due to his love of horses and riding.

At the start of the war Manfred served in the cavalry on both the Eastern and Western Fronts and survived a number of difficult combats. Once the Western Front settled into trench warfare, the cavalry fought as infantry, and like many cavalrymen Manfred was bored by this. Early in 1915 he requested a transfer to aviation and was accepted for observer training. This path was typical for many cavalrymen because they were already trained in reconnaissance on horseback. He flew as an observer and later volunteered for pilot training. He was flying as a pilot in Russia when Oswald Boelcke visited his unit. Boelcke was choosing pilots for Germany's first fighter unit and asked Manfred to volunteer – an opportunity he immediately grasped.

Soon Manfred was one of the fledgeling fighter pilots in Boelcke's *Jasta* 2 where he received the same exceptional coaching and leadership from

Above: Manfred von Richthofen (with binoculars) and fellow *Jasta* 2 pilots in front of von Richthofen's Albatros D.II fighter. Known most for the Fokker Triplane he was flying when killed, he scored most of his victories in Albatros fighters.

Boelcke as the other new pilots. Manfred revered Boelcke and took his lessons and methods to heart, shooting down his first enemy aircraft during his first encounter with the enemy.

An apt pupil, Manfred soon scored more victories, and by January 1917 was promoted to command *Jasta* 11 and was awarded the *Pour le Mérite*. *Jasta* 11 had not yet scored a single victory, so Manfred led by example and scored a victory his first day at the new unit. He scored another on his second day at the unit, and under his tutelage and following his example, *Jasta* 11 quickly became the highest-scoring *Jasta* in the *Luftstreitkrafte*.

At *Jasta* 11, Manfred first began to paint his aircraft red. He first tried various camouflage schemes, but when none were effective under all the varied flight conditions, he turned to painting his

Right: Manfred von Richthofen (left) and his younger brother Lothar in heavy flying clothing.

Above & Above Right: Manfred and his dog Moritz in lighter moments showing a side of Manfred not normally mentioned.

aircraft red so the other *Jasta* 11 pilots could easily see him. He did not paint his aircraft red to flaunt himself in front of his enemies as has been claimed.

March 1917 was a big month for Manfred in a number of ways. On the 6th his younger brother Lothar joined *Jasta* 11, and on the 23rd he was promoted *Oberleutnant*. This was followed just two weeks later by another promotion to *Rittmeister*! April 1917 would go down in RFC history as "Bloody April" and *Jasta* 11, led by Manfred, was a major contributor to the RFC's misery. *Jasta* 11 accounted for 89 victories in April, with Manfred scoring 21 of them, Lother scoring 15, Kurt Wolff surpassing his CO's count with 22, Karl-Emil Schaefer with 15, and Sebastian Festner with 10 among the big scorers.

As a result of their success, *Jasta* 11 was honored by a visit from *GenLt*. Ernst von Hoeppner, commanding general of the German Air Service, on 23 April to personally congratulate them on their 100th overall victory. There was also widespread newspaper coverage of the men complete with many photographs taken at their airfield. The aces of *Jasta* 11 quickly became national heroes, and postcards featuring their photographs were sold across Germany. Moreover, newspapers began to follow their exploits on an almost daily basis.

As their leader, Manfred was the most famous of them all. His mother noted in her diary:

"He's in all the newspapers, everyone is talking about him; a banner waves above his name. Cities honor him, royalty send telegraphs. No sooner have the congratulations arrived when a new victory chases the flag up the flagpole. It is unbelievable. The enemy is completely flustered."

On June 24, Manfred became the leader of *JG* 1 (*Jagdgeschwader* 1), Germany's first fighter wing, composed of *Jastas* 4, 6, 10, and 11. This had been formed in response to the larger fighter formations the Allies had started employing. There was little

Above: Manfred von Richthofen was survived by his younger brother Lothar, who was also a successful fighter pilot. Starting later than Manfred, Lothar scored 40 confirmed victories and, like Manfred, was awarded the Blue Max. Lothar had a less disciplined fighting style than Manfred which resulted in him being wounded several times, but he survived the war with 40 victories.

Above: A replication, using authentic period decorations, of Manfred von Richthofen's *Ordenskissen* as displayed at the Richthofen Museum in Schweidnitz. Once, upon being praised, Manfred von Richthofen replied, "I am but a combat pilot, but Boelcke, he was a hero!"

time to celebrate; on July 6 Manfred was wounded in combat with a formation of FE.2d aircraft of No.20 Squadron. Manfred was hit a glancing blow to the head by shots fired from 300 meters and was nearly killed. He returned to his unit some weeks later and scored his next victory on August 16, but had not still fully recovered from his near-fatal wound.

Nevertheless, Manfred continued to do his duty. He had instigated the fighter competitions and flew prototypes at the first one in January 1918, helping select Germany's new fighter, the Fokker D.VII. He was not to use it in combat, being killed just before it arrived at the front. On April 21, 1918, he broke one of his own rules of fighting and followed a Sopwith Camel low over enemy territory, where hundreds of soldiers had clear shots at his plane. He was fatally wounded, almost certainly by ground fire, and buried with full honors by his faithful enemies in the Royal Flying Corps.

Right: Manfred being helped into his flying clothes.

Above: This photo of a typical German *Jasta* in 1918 shows the variety of aircraft common to the units due to inadequate supplies of the preferred Fokker D.VII; from left, Fokker D.VII, Albatros D.Va, Fokker D.VII, Fokker Dr.I triplanes.

The "Flying Circus" or "Richthofen's Circus" or "The Red Baron and his Flying Circus" are all names for one of the legends of the First World War. What is the truth behind the legend?

The real "Flying Circus" was officially known as *Jagdgeschwader Nr.*I, or *JG* I for short, and was a permanent combination of *Jastas* 4, 6, 10, and 11 officially established on June 24, 1917. It was the first fighter wing in the *Luftstreitkräfte* (German Air Service) and its first commander was *Rittmeister* Manfred von Richthofen, the Red Baron, who had commanded *Jasta* 11. *JG* I was a response to the increasingly larger Allied formations the German fighters were meeting in combat, forcing the German fighters to combine into larger units. Because the German Air Service was out-numbered, *JG* I was moved back and forth along the Western Front to wherever local air superiority was most important.

There was no single reason this unit, known to most German airmen as '*Geschwader* Richthofen,' was called the "Flying Circus" by Allied airmen. One reason was that *JG* I moved up and down the front, often transported on trains, like travelling circuses of the time. Another was the gaudy colors of the aircraft, a riot of color not seen on British aircraft of the time. The variety of fighter types used by German units no doubt contributed to the spectacle. Less obvious reasons included the habit of calling any apparently chaotic movement of people or machines a 'circus', and using 'circus' as an under-stated nickname for a very tough, dangerous, disciplined, and effective aerial opponent.

Richthofen's Circus made an immediate impact wherever it appeared and earned great respect from friends and foes alike. *JG* I was so successful that three more permanent *Jagdgeschwaders* were formed. The next two *Jagdgeschwaders, JG* II and *JG* III, were formed on February 2, 1918, in preparation for Operation Michael, or the *Kaiserschlacht*, the German's 1918 Spring Offensive that they hoped would win the war before significant numbers of American troops arrived on the Western front. *JG* II was formed from *Jastas* 12, 13, 15, and 19, and *JG* III was formed from *Jastas* 2 (Boelcke), 26, 27, and 36. This enabled a *Jagdgeschwader* to be assigned to each of the three attacking armies.

Finally, on October 3, 1918 *JG* IV, a Bavarian *Jagdgeschwader*, was formed from Bavarian *Jastas* 23b, 32b, 34b, and 35b in recognition of Bavaria's status as the second most powerful state in Imperial Germany, behind only Prussia.

The main limitation preventing more *Jagdgeschwaders* from being formed was the lack of enough highly effective fighter leaders. Instead of forming more permanent *Jagdgeschwaders* led by less qualified officer pilots, a system of temporary groupings of *Jastas* into *Jagdgruppes* was initiated in October 1917. Typically two *Jastas* were grouped into a *Jagdgruppe* based on the tactical situation. In contrast to the *Jagdgeschwaders,* which were led by fighter aces in the air, the *Jagdgruppes* were generally directed by orders from the ground by

Above: Lineup of Albatros D.V fighters of *Jasta* 11 in the summer of 1917. Red was the unit color of *Jasta* 11 and these fighters all have red noses, struts, and wheel covers. The tails were in the personal colors of the individual pilots, enabling them to be distinguished from one another during combat.

a *Führer der Jagdgruppe* (*Gruja*). As a result, the *Jagdgeschwaders* were normally more effective in combat than the *Jagdgruppes*.

It should be noted that the French had initiated combining several fighter squadrons into *Groupes de Combat* months before the first "Flying Circus" was established, just as they had formed the first dedicated fighter squadrons months before Germany did the same.

Below: Lineup of Pfalz D.IIIa and Albatros fighters of *JG* I at *Flugplatz* Awoingt, March 1918. The Pfalz fighters in the foreground were assigned to *Jasta* 10 and the fighters in the background were assigned to *Jasta* 4, two of the four *Jastas* assigned to *JG* I. *Jastas* 6 and 11, the latter formerly commanded by Manfred von Richthofen, were the other two *Jastas* of *JG* I.

Above: The extensive American flight instruction complex at Issoudon, France, was the largest in the world. This is the 3rd Aviation Instruction Center, one of *nine* A.I.C.s at Issoudon.

Below: Nieuports at Field #5 are precisely aligned for inspection in mid-March 1918. Ship #690 N.2587 is a typical Type 21 with 80 hp Le Rhône. The sloping ammunition feed reveals it was formerly a Type 17. It was delivered to 3 AIC on 16 February 1918. But next in line is ship #3, one of the first to become operational at 3 AIC in October 1917. Its lack of a headrest suggests it was constructed using a Type 11 fuselage, but now fitted with 15m^2 Type 17 wings and a circular cowl. The third ship has a cowl with panel on top, suggesting it was formerly a Type 23. These aircraft were used to train American pilots in France. (*Peter M. Bowers collection via Alan Toelle*)

Above: Eddie Rickenbacker, the leading American ace of the war, is shown in his Nieuport 28 early in his fighting career.

Left: Rickenbacker after the partial upper-wing failure of his Nieuport 28. The Nieuport 28 was a French design that only saw combat with the USAS. An elegant, fast, and maneuverable fighter, it was initially liked by its pilots. However, a number of incidents like that shown here, together with a propensity to catch fire in flight due to broken fuel lines, soon caused the pilots to demand a better fighter. The initial American request for the Spad 13 had been denied due to insufficient production and the Nieuport 28 was substituted, but the Nieuport's spectacular failures and increased Spad production soon lead to the Nieuport's replacement by the more robust and reliable Spad – to the great relief of the pilots. Plans were made to re-introduce the Nieuport 28 after its problems had been fixed, but the war ended before that could happen.

Above: Nieuport 28s of the 95th Aero Squadron about to take off. Most engines are running and the two men in the foreground are preparing to pull the wheel chocks.

Below: Nieuport 28 displayed at the National Air & Space Museum in the markings of the 94th Aero Squadron.

Above: Officially credited with 26 victories, Capt. Eddie Rickenbacker was America's leading WWI ace and one of eight airmen awarded the Medal of Honor. Here he is seen with his Spad 13 while serving in the 94th Aero Squadron.

Eddie Rickhenbacher was born to German-speaking Swiss immigrants in Ohio, Before America's entry into WWI he was a successful race car driver and mechanic. During the war, with its anti-German atmosphere, he — like many other German Americans — changed his surname; the "h" in "Rickenbacher" became a "k" in an effort to "take the Hun out of his name." He was already well known at the time, so the change received wide publicity.

Once America entered the war Rickenbacker enlisted in the US Army and was soon sent to France, arriving there as a Sergeant First Class. He was assigned as engineering officer at the 3rd Aviation Instruction Center at Issoudun, the US Air Service's pursuit training facility, where he learned to fly during his spare time. Once he demonstrated he had a qualified replacement, he was assigned to the 94th Aero Squadron, one of the first American squadrons to reach combat.

Rickenbacker first flew the elegant Nieuport 28 and claimed his first victory on April 29, 1918. On May 28th he became an ace after claiming his fifth victory, and claimed a sixth two days later. In July he experienced an ear infection that nearly ended his flying career. Once recovered, he continued scoring, now flying the robust Spad 13. On September 24 he was made commanding officer of the 94th Aero Squadron. Five of his victories were balloons and 13 were Fokker D.VIIs, Germany's best fighter.

Postwar Rickenbacker started the Rickenbacker Motor Company in 1920, and in 1927 bought the Indianapolis Motor Speedway, which he operated until 1941. He also had a business relationship with General Motors, which asked him to manage Eastern Air Transport in 1935. He merged that company with Florida Airways to form Eastern Airlines, and in 1938 bought the airline from GM. As a passenger, Rickenbacker survived a DC-3 crash in 1941 that would have killed a less determined man. In October 1942, while on a tour of Pacific air bases, the B-17D he was in was forced to ditch due to navigation error and he survived a 24-day ordeal in a life raft. Eastern Airlines initially prospered after WW2, but in 1959 Rickenbacker was forced out after financial reverses. He died July 23, 1973 while in Zurich, Switzerland.

The German Fighter Competitions

For a number of reasons – technical stagnation, partly due to engine development stagnation, increasing Allied technical and numerical superiority, and American entry into the war making all these problems worse – German fighter aviation was under great stress in the second half of 1917.

At this time, Manfred von Richthofen, the 'Red Baron', Germany's leading ace famous throughout the Empire, contacted his friend *Oblt.* Fritz von Falkenhayn, (son of *Gen.* Erich von Falkenhayn, Chief of the General Staff from Sept. 1914 to August 1916) who was *Kogenluft's* adjutant 1915–1918, and complained strongly about 'the wretched Albatros' fighters. Richthofen suggested a fighter competition between all the German manufacturers to develop a superior new fighter. Such was his influence, and the common sense of the solution he recommended, that *Idflieg* decided to follow his suggestion, and the first German fighter competition was held 20 January–12 February 1918.

Above: Manfred von Richthofen, the famous Red Baron, was not just the war's leading ace and fighter leader. He also served his country by instigating the fighter competitions held three times during 1918 that were instrumental in getting new fighter types to the front.

All the manufacturers entertained the pilots and officials to make them well-disposed to their aircraft. However, *Idflieg* had given a lot of thought to the fighter competition process and designed it to be as fair and objective as possible and to address the priorities of successful fighter pilots at the front.

To ensure unbiased results, strict procedures were instituted. Once the prototypes were handed over to the Adlershof staff, they were basically in bond, and no modifications or adjustments were to be made by the manufacturers. The prototypes were checked and weighed before each flight. Details of the propellers fitted were recorded, and the fuel consumed during each flight was recorded. Two calibrated and sealed barographs were fitted to each aircraft; these were checked before and after every flight.

At the beginning of the trials the aircraft were demonstrated by company pilots under supervision of the Adlershof staff. The second part of the trials were flown by Adlershof pilots and front-line pilots from the *Jastas*, especially well-known, successful aces. The front-line pilots were to fly the prototypes for the key part of the trials, including climb and speed performance, maneuverability, and simulated combat with other prototypes and service aircraft provided for comparison. These aces were to evaluate the competitors and make recommendations about which designs should actually be produced. In addition to testing flight performance, the aircraft were evaluated for strength, reliability, and maintainability.

Moreover, examples of the fighters currently at the front, or more powerful versions of them, were included in all the competitions to provide a practical standard of comparison for the competitors.

The key result of the First Fighter Competition, whose formal trials were held from 21 January until 12 February, 1918, was the selection of a superior fighter, the Fokker D.VII, to replace the outdated Albatros and Pfalz fighters. Significantly, the D.VII was initially powered by the same Mercedes engine used in those fighters, but soon a new engine, the over-compressed 185 hp BMW.IIIa became available, enabling more of the airframe's potential to be exploited. Mercedes responded with improved versions of their basic engine offering similar performance. The resulting BMW-powered D.VII

Above: The most important immediate result of the First Fighter Competition, and the most important result of all the fighter competitions, was the selection of the Fokker D.VII for large-scale production for the German fighter units. When fitted with a BMW engine this fighter regained technical superiority for Germany and enabled her pilots to double their scoring rate. This is an original D.VII in its original camouflage now on display in Knowlton, Canada.

fighter is thought by many, including the author, to be the finest all-around fighter of the war.

The other important result of the First Fighter Competition was that it showed the efficacy of the process in procuring new, superior fighters for the air service. In general, the fighter competitions provided more structure, focus, and momentum to the German effort to develop and procure better fighters for their air service. Therefore, two subsequent fighter competitions were held.

In mid May all manufacturers were notified that their prototypes had to be at Adlershof by 31 May. The aircraft were first inspected, weighed, and prepared for testing. Performance testing was done June 1–21, followed by comparative evaluation.

The Second Fighter Competition resulted in production of the Fokker E.V/D.VIII, a parasol-monoplane development of the Fokker Dr.I and D.VI. This innovative aircraft had a troubled history due to manufacturing quality problems, but the design itself was sound and offered excellent performance considering its simplicity and low power.

The Third Fighter Competition, officially restricted to fighters with BMW.IIIa engines but including rotary-powered types for comparison, was held October 15–31, 1918. The Fokker V.29, a parasol-monoplane derivative of the Fokker D.VII, was one of two winners and may have gone into production except for the Armistice. The Rumpler D.I, also chosen as a winner, was still not ready for production before the Armistice despite a prolonged development process that had started in early 1917. Whether the troubled Rumpler D.I would ever have been brought to production-ready status as a practical fighter is doubtful.

Given the results of the competitions, some critics have contended that Anthony Fokker had an unfair advantage. Certainly Fokker himself was an exceptional pilot, and he also had great rapport with many of the fighter pilots evaluating the prototypes at the competitions. This clearly gave him greater insight into the real needs of the combat pilots at the front than his competitors, and his abilities as a test pilot helped him fine-tune his aircraft to meet the needs of the aces. These were certainly advantages, but not unfair ones; Fokker's aircraft still had to deliver better performance than those of the other manufacturers to win the competitions. Moreover, from the sheer number of different prototypes he brought to the competitions, it is apparent that Fokker simply worked harder to succeed than the other manufacturers.

Fokker had struggled since the eclipse of his *Eindeckers*, and his famous Triplane was only a partial success. Limited to the same engines and weapons his competitors used, Fokker needed a significant breakthrough to build a dramatically better airplane, and he finally achieved it in the Fokker D.VII with its wing of innovative structural and aerodynamic design.

Above: One of the greatest services Manfred von Richthofen did for Germany was instigating the fighter competitions, and the Fokker D.VII was the greatest result of those competitions. Arriving at the front days after Richthofen's death in his Fokker Triplane, he was not able to fly the D.VII in combat himself. Initially using the 170 hp Mercedes D.IIIa engine, when fitted with the superb 185 hp BMW.IIIa it was the best all-around fighter of the war. This D.VII flew with *Jasta* 49.

Although sharing its engine and armament with the Albatros, Pfalz, and other German designs, the Fokker D.VII introduced important structural and aerodynamic innovations that greatly improved its effectiveness. By far the most important was its thick wooden wing built around a box spar. The thick wing, with its rounded leading edge, offered high lift and exceptional stalling characteristics, making the D.VII maneuverable and easy to fly and enabling it to 'hang on its prop' without stalling. These exceptional handling qualities made good pilots out of average ones and made aces out of good pilots. Other designers used thin airfoils because they had somewhat less drag than the thick Fokker wing. However, the strong Fokker wing eliminated the need for the extensive system of bracing wires that thin airfoils required. The combined drag of thin airfoils with their bracing wires was significantly more than the drag of the thick Fokker wing which needed no bracing wires, and this was the secret to the Fokker's improved speed and climb with the same power.

Due to stagnant engine development, upon its introduction to service the Fokker D.VII was powered by the same basic Mercedes engine used in the Albatros scouts in August 1916! It was not until the new BMW engine finally arrived in June/July that the Fokker D.VII fulfilled its full potential and became the premier fighter of the war. The BMW engine was similar to the familiar 160 hp Mercedes D.III engine but developed its 185 hp at 2,000 meters altitude because it was over-compressed. That meant it could not be run at full throttle until reaching the thinner air at 2,000 meters without detonation and engine damage. This design gave it more power at high altitude for increased speed and exceptional climb. Mercedes countered with the over-compressed 180 hp D.IIIaü and finally the 220 hp D.IIIavü, the latter having greater bore. Both engines were used in the D.VII, with the D.IIIavü giving performance equivalent to the BMW. Although the BMW-powered Fokker D.VII became a legend in its own time, there were never enough of them to win the air war for Germany.

Above: Reproduction Fokker D.VII displayed in authentic camouflage and markings at the USAF Museum. This D.VII is in the colors of *Lt*. Rudolf Stark, *Jastaführer* of Jasta 35b ('b' indicating a Bavarian unit), 11 victory-ace, and author of the postwar book ***Wings of War***.

Below: Reproduction D.VII in authentic camouflage and representative markings evokes the first war in the air.

Fokker D.VII Specifications

Engine:	185 hp BMW.IIIa	
Wing:	Span	8.90 m
	Area	20.2 m^2
General:	Length	6.954 m
	Height	2.945 m
	Empty Weight	688 kg
	Loaded Weight	906 kg
Maximum Speed:		200 km/h
Climb:	1000m	1.8 min
	2000m	4.0 min
	3000m	7.0 min
	4000m	10.2 min
	5000m	14.0 min
	6000m	18.7 min

Germany's Fighter Competitions of 1918			
Competition	**Dates**	**Winners**	**Remarks**
1st	20 Jan.–12 Feb.	Fokker D.VII biplane	Evaluation batches of other types also ordered[1]
2nd	27 May–28 June	Fokker D.VIII monoplane	Evaluation flights by front-line pilots in July[2]
3rd	Oct. 10–Nov. 2	Fokker V.29 monoplane Rumpler D.I biplane	8 monoplane and 7 biplane competitors[3]

Notes:
1. 50 Roland D.VI ordered, 120 Fokker D.VI ordered (later reduced to 60), Pfalz D.VII/D.VIII ordered, SSW D.III ordered, 50 Rumpler D (7D4) ordered. Only one *new* triplane type was entered.
2. 50 Pflaz D.XIV ordered after second competition; most apparently cancelled. One quadraplane (Naglo D.II) was entered and no triplanes, indicating the 'triplane craze' was finally over.
3. Officially, the third competition was for BMW-powered fighters only, but some rotary-powered fighters were entered for comparison. An evaluation batch of the all-metal Zeppelin D.I was ordered.

Left: The forbidding continence of *Lt.* Jakob Wolff, one of the oldest German fighter pilots, who was 46 when the war started. Wolff looks straight out of Hollywood's central casting as a typical German ace, and is a distinct contrast to the actual Red Baron as seen at play with his dog – although he does bear some similarity to Manfred in the leading photo. Wolff has his own interesting story; at the start of the war, he volunteered for the air service in spite his age and position as a wealthy business owner. No one in his situation was expected to fight at the front. Despite paying for his own flying lessons, his age and Jewish religion worked against his acceptance as a combat pilot, but his persistence and ability eventually enabled him to achieve his goal. Joining *Jasta* 17, Wolff claimed seven victories, four of them confirmed, and won a number of awards for bravery and leadership before being severely wounded in combat. He survived, but his flying career was ended and he was invalided out of the army.

Above: Pilots of *Jasta* 4 in front of a Fokker D.VII of their unit, 5 September 1918. The Fokker D.VII became the main German fighter as soon as its production could re-equip the *Jastas*.

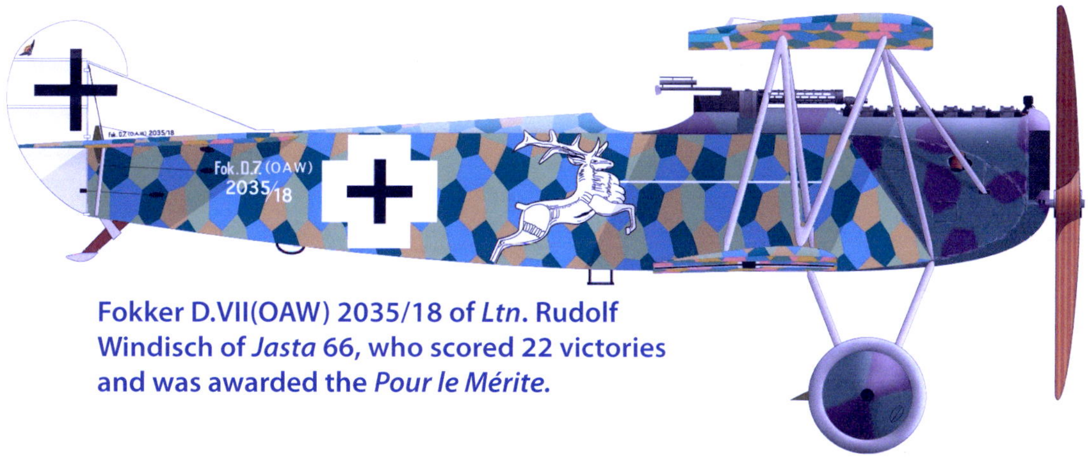

Fokker D.VII(OAW) 2035/18 of *Ltn*. Rudolf Windisch of *Jasta* 66, who scored 22 victories and was awarded the *Pour le Mérite*.

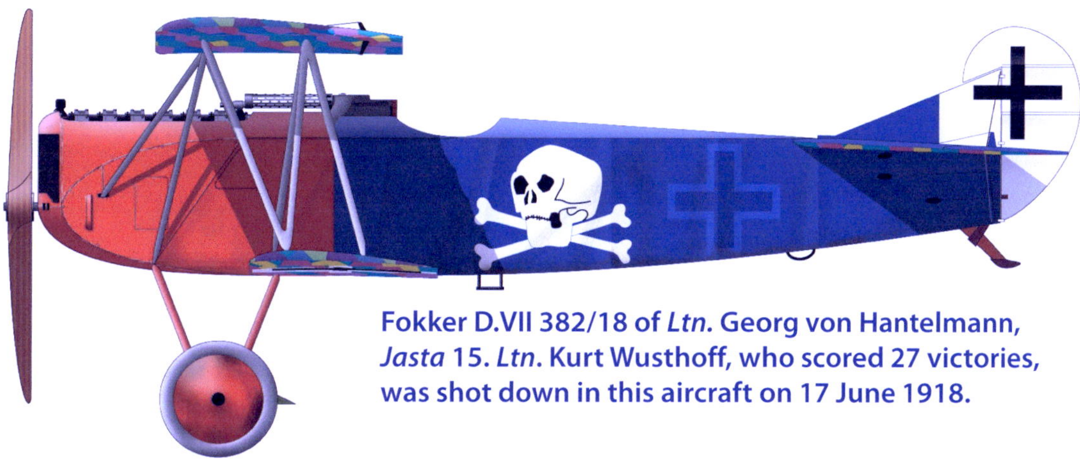

Fokker D.VII 382/18 of *Ltn*. Georg von Hantelmann, *Jasta* 15. *Ltn*. Kurt Wusthoff, who scored 27 victories, was shot down in this aircraft on 17 June 1918.

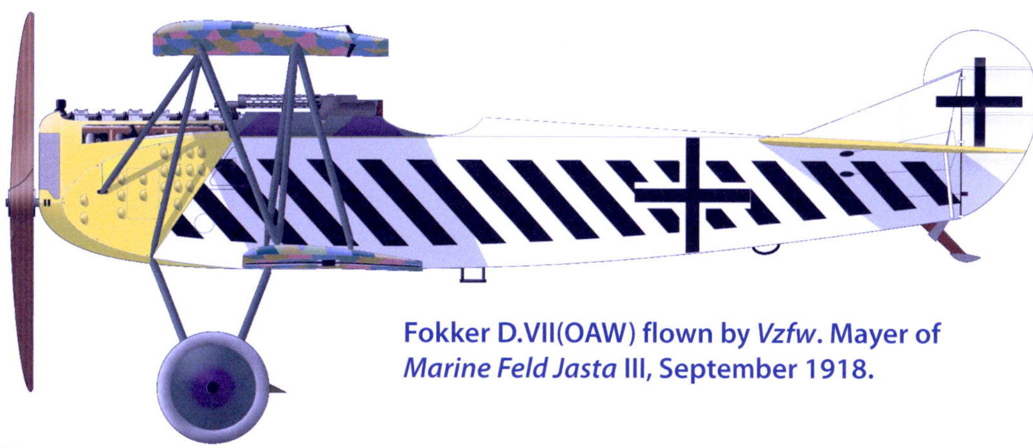

Fokker D.VII(OAW) flown by *Vzfw*. Mayer of *Marine Feld Jasta* III, September 1918.

Fokker D.VII(Alb) of *Ltn*. August Raben, *Jastaführer* of *Jasta* 18, Montingen Aerodrom, August 1918. Raben scored 4 victories.

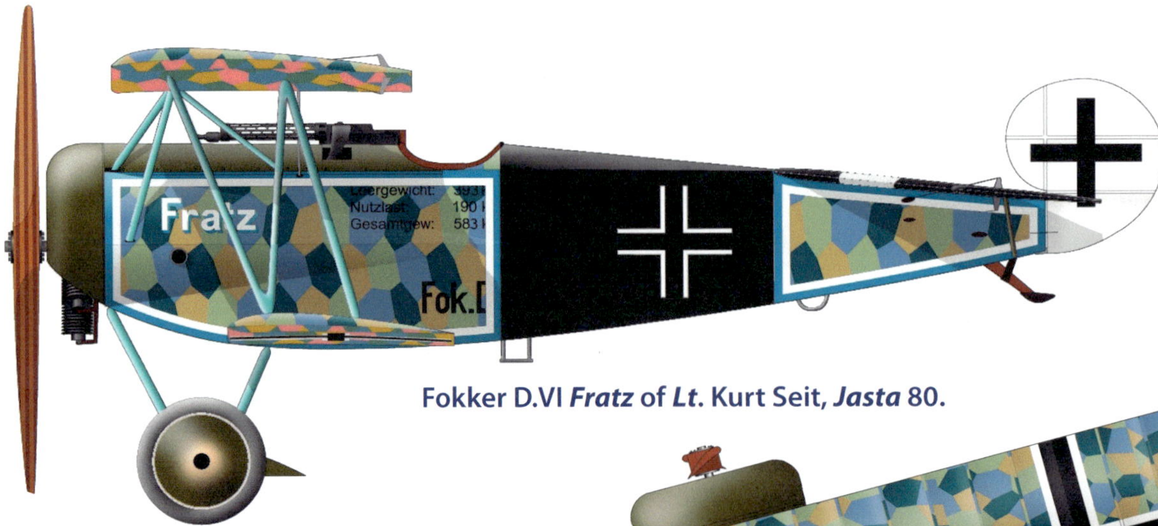

Fokker D.VI ***Fratz*** of ***Lt.*** Kurt Seit, ***Jasta*** 80.

Fokker D.VI of ***Oblt.*** Max Speidel, ***Jasta*** 80.

Right: Fokker D.VI of *Oblt*. Max Speidel of *Jasta* 80b. The center fuselage was painted black (or, because it was a Bavarian unit, possibly dark blue) with white stripes, and white stripes were added on the tailplane and between the crosses on the top of the upper wing.

A production batch of 120 Fokker D.VI fighters was ordered after the First Fighter Competition. The little-known D.VI clearly shows the evolution of Fokker fighters from the Dr.I triplane though the D.VI biplane, D.VII biplane, and D.VIII parasol monoplane. Essentially, the D.VI was a monoplane derivative of the Dr.I triplane and used the same low-powered rotary engine, the nine-cylinder, 110 hp Oberursel Ur.II, for lack of a more powerful alternative. The 11-cylinder, 145 hp Oberursel Ur.III was in development but not yet ready for production.

The D.VI was an excellent airframe that had the misfortune to be under-powered. It was smaller, lighter, and more maneuverable than the D.VII and was faster at low altitude, but the Mercedes-powered D.VII had better high-altitude performance, so important by 1918, and was chosen for mass production instead.

Production of the D.VI was cut to 60 aircraft and the order for the last 60 was transferred to the D.VIII monoplane after the Second Fighter Competition. The D.VIII also retained the 110 hp Oberursel Ur.II until the last batch; delivered on 8 October, this batch of D.VIII fighters finally had the 145 hp Oberursel Ur.III so desperately needed by the Fokker rotary-powered fighters.

Below: D.VI 1689/18, the last production aircraft, photographed 21 June 1918. The D.VI had excellent visibility for the pilot, was very maneuverable, and was light and sensitive on the controls, a real pilot's airplane. It just needed more power than its 110 hp Oberursel Ur.II provided to be a first-class fighter.

Above: The Fokker V17 participated in the First Fighter Competition powered by a 110 hp Oberursel Ur.II. Officially rejected due to inadequate field of view downward, the preference for sturdy biplanes may also have influenced the decision. The Fokker D.VI biplane powered by the same engine as the V17 was selected for limited production after competing with the V17 at the First Fighter Competition despite the V17 offering better speed and climb rate. It is interesting to speculate on the combat success of the V17 had it been produced in quantity for operational service, which could have happened five or six months prior to the appearance of the later Fokker E.V that was still powered by the same low-power engine. By its rejection of the V17 *Idflieg* may have missed an excellent opportunity. Inability to bring the 11-cylinder, 145 hp Ur.III rotary into timely production was also a major impediment to all of Fokker's 1918 rotary fighter designs, which needed more power. The Ur.II had only ten more horsepower than the U.I of 1915! The first acceptance of a production fighter with the 145 hp Ur.III, a Fokker D.VIII, was October 8, 1918, a mere month before the Armistice.

Below: The Fokker V20 was the Mercedes-powered sibling of the Fokker V17. Like the V17, it was officially rejected due to inadequate field of view downward, and probably due to skepticism about monoplanes.

Above: Like the similar Fokker V20, the Fokker V23 was rejected because of the poor downward view for the pilot due to the wing location. Power was a 160 hp Mercedes D.III.

Above & Below: The small Fokker V25 with 110 hp Oberursel Ur.II engine.

Above: The Fokker V27 parasol monoplane configuration solved the problem of the pilot's limited field of view suffered by the previous Fokker monoplane prototypes at the expense of greater drag from the struts supporting its wing. However, despite good performance its 195 hp Benz Bz.IIIbo experimental V-8 was not ready for production.

Above: The Fokker V28 parasol monoplane configuration solved the problem of the pilot's limited field of view suffered by the previous Fokker monoplane prototypes at the expense of greater drag from the struts supporting the wing. Smaller and lighter than its companion V-8 powered V27, the V28 was the prototype of the production Fokker E.V that went into production with the 110 hp Oberursel Ur.II previously used in the Fokker Triplane and Fokker D.VI biplane. Here the V28 is seen with the 11-cylinder Ur.III of 145 hp.

The Fokker V28 was the winner of the Second Fighter Competition and was placed into production as the Fokker E.V, unfortunately with the same low-power 110 hp Oberursel Ur.II used previously in the Fokker Triplane and Fokker D.VI.

The E.V was the minimum practical fighter aircraft that could be built with the available technology. It was strong, fast, and maneuverable, although its cantilever wings vibrated during high-G maneuvers and its aileron response was not as good as the Fokker biplanes with their torsionally stiffer wing cellules.

Production Fokker E.V fighters started arriving at the *Jastas* by 5 August 1918. Unfortunately, fatal in-flight wing failures followed on 16 and 19 August, and the E.V was immediately grounded.

The problem was traced to rot of the inner wing structure due to moisture condensation in the wing. The moisture had entered the wing via breathing holes and condensed. The problem was compounded by poor production quality control at the factory, which resulted in under-sized spars and faulty production procedures.

As a result, a new wing with strengthened spars was tested on 7 September and passed the structural tests. On 24 September Fokker was given permission to resume building the fighter, now re-designated the Fokker D.VIII. Fokker was also required to build new wings at his own expense for the 139 E.V fighters already completed. By the Armistice 80 D.VIII fighters had been accepted, but it is not known how many of these reached combat.

The last batch of Fokker D.VIII fighters were ordered with the 11-cylinder, 145 hp Oberursel Ur.III. Some sources give the Ur.III power as 160 hp but 145 hp is more likely given the Ur.III was an 11-cylinder derivative of the 9-cylinder/110 hp Ur.II.

The results of the Third Fighter Competition showed that better fighters were available, and the D.VIII was likely destined for a relatively short career even had the war lasted into 1919.

Fokker E.V / D.VIII Specifications

Engine:	110 hp Oberursel Ur.II	
Wing:	Span	8.34 m
	Area	10.7 m^2
General:	Length	5.865 m
	Height	2.82 m
	Empty Weight	405 kg
	Loaded Weight	605 kg
Maximum Speed:		200 km/h
Climb:	1000 m	2.0 min.
	2000 m	4.5 min.
	3000 m	7.5 min.
	4000 m	10.8 min.
	5000 m	15.0 min.
	6000 m	19.5 min.

Above: The production Fokker E.V, powered by the 110 hp Oberursel Ur.II previously used in the Fokker Triplane and Fokker D.VI biplane, in service with *Jasta* 6, August 1918. The wing tips were more rounded than on the V28. Like most Fokker fighters the E.V was a responsive, maneuverable 'pilot's airplane' with good handling qualities. Built as designed the wing was very strong, but poor production quality control resulted in wing failures and immediate removal of the E.V from the front. On Oct. 8, 1918, the first production D.VIII fighters powered by the 11-cylinder, 145 hp Oberursel Ur.III were delivered. Mere days before the Armistice, the D.VIII finally had the engine it (and the earlier Dr.I and D.VI) needed.

Fokker E.V of *Off.Stv*. Friedrich Altemeier, *Jasta* 24. Altemeier scored 21 victories while flying various fighter types.

Fokker E.V 149/18 of *Lt*. Friedrich-Wilhelm Liebig, *Jasta* 1.

Fokker E.V of *Jasta* 36.

Fokker E.V of *Vzfm*. Hans Goerth, *Marine Feld Jasta* 3. Goerth scored 7 victories flying various fighter types.

After winning the first two fighter competitions, Fokker entered the Third Fighter Competition in the best position of any manufacturer. The Fokker D.VII was the mainstay of the German fighter units and the BMW-powered version was generally acknowledged as the best all-around fighter in combat. Fokker's D.VII was being built not only by Fokker but by other manufacturers under license. After a rocky start due to two fatal wing failures, Fokker D.VIII production was back in full swing.

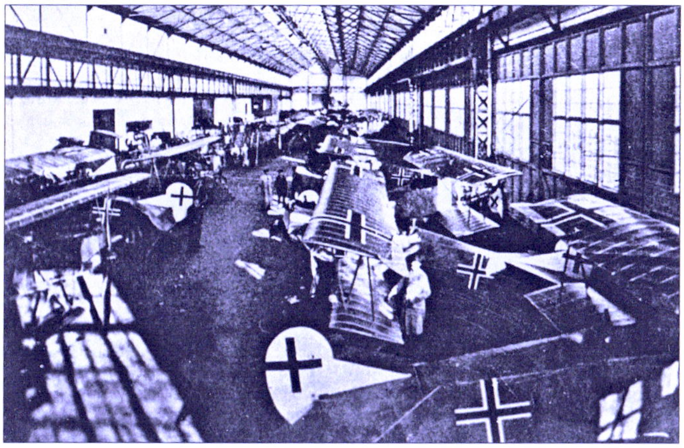

Above: Fokker D.VII mass production at Fokker.

Fokker's entries in this competition built on these successes. The V28 was a D.VIII prototype with the more powerful Seimens-Halske rotary, the V36 was a modified D.VII, and the V29 was a parasol monoplane derivative of the D.VII. Using the same basic configuration as the production D.VIII, the V29 depended on its over-compressed BMW engine for excellent high-altitude performance and was one of the winners of the competition. However, with fresh memories of the D.VIII wing debacle and with the robust, proven D.VII still superior, *Idflieg* was in no hurry to order the V29 into production.

Fokker V29 Specifications

Engine:	185 hp BMW.IIIa	
Wing:	Span	9.68 m
	Area	14.3 m^2
General:	Length	7.02 m
	Height	2.93 m
	Empty Weight	632 kg
	Loaded Weight	861 kg
Maximum Speed:		200 km/h
Climb:	3000 m	7.5 min.
	4000 m	10.5 min.
	5000 m	14.7 min.
	6000 m	19.7 min.

Below: The Fokker V29 shown here was one of the two winners of the Third Fighter Competition. Essentially, it was a parasol monoplane derivative of the Fokker D.VII. Similar to the smaller, D.VIII, it had the more powerful 185 hp BMW.IIIa engine for better high altitude performance. Similarly, the more powerful D.VII biplane had eclipsed the smaller, rotary-powered D.VI of similar configuration.

This Page: Based on its showing at the First Fighter Competition, the SSW D.III was placed in limited production for evaluation. SSW D.III 8341/17, the second production D.III fighter is shown at the SSW factory at Siemensstadt. Its full cowling over its Sh.III rotary and camouflage printed fabric on the wings and movable tail surfaces are clearly shown.

SSW D.III 8341/17, the second production D.III, was sent to *JG* III on March 16, 1918.

SSW D.III 8346/17 of ***Lt.*** Walter Göttsch, *Jasta* 19, April 1918. Goettsch was KIA in his Fokker Triplane while scoring his 20th victory.

SSW D.III 8342/17 of *JG* III.

SSW D.III of *Kest* 4b.

Above & Below: SSW D.IV 3082/18 of *JG* II photographed October 25, 1918. Powered by a 205 hp Sh.IIIa counter-rotary, the SSW D.III was maneuverable and had an exceptional climb rate, but – as always! – pilots wanted a faster fighter. SSW responded with the D.IV, which was essentially a D.III modified by reducing the chord and area of the upper wing. The reduced weight and drag was intended to boost speed at the expensive of a small reduction in climb rate. Speed was increased but climb rate seemed unaffected; perhaps the reduced weight and drag compensated for the smaller wing area. In any case, pilots found the D.IV could out-climb and out-maneuver any other fighter at the front in late 1918 and considered it the best fighter of the war, an honor more frequently bestowed on the BMW-powered Fokker D.VII. The next logical step in development was a parasol monoplane, and the SSW D.VI was much faster still, but too late for combat.

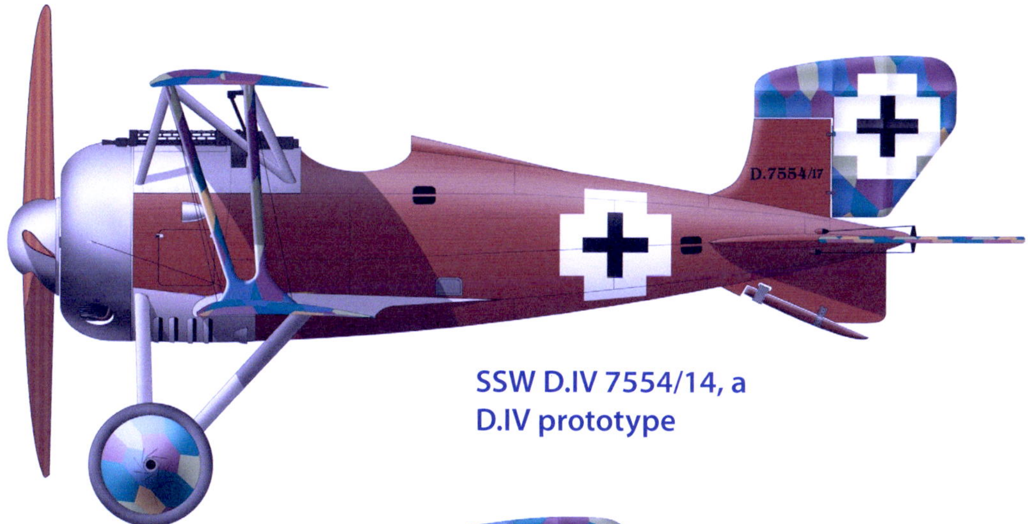

SSW D.IV 7554/14, a D.IV prototype

SSW D.IV 7555/14, a D.IV prototype

SSW D.IV of *Jasta* 12

SSW D.IV of II. *Marine Feld Jasta* flown by *Lt.z.S.* S. Franz

Pfalz D.III of *Vzfw.* Friedrich Rüdenberg, *Jasta* 10

Pfalz D.IIIa of *Vzfw*. Max Holtzem, *Jasta* 16b

Pfalz D.VIII of *Lt*. Paul Bäumer, *Jasta* Boelcke

Pfalz D.XII 1394/18, *Jasta* 77b. The rectangular fin and rudder were characteristics of early production aircraft.

This Page: The Pfalz D.VII made a good impression at the First Fighter Competition. After testing its stronger, two-bay version the D.VIII was put in limited production. *Lt*. Paul Bäumer of *Jasta* Boelcke and his Pfalz D.VIII are shown here. Bäumer scored three victories with *Jasta* 5, then transferred to *Jasta* Boelcke (honor name for *Jasta* 2) in August 1917 and scored 40 more victories to bring his total to 43. Bäumer was injured in a crash of a Pfalz D.VIII on 29 May 1918 and did not return to his unit until September. He then scored 16 victories in 25 days! Known as *Der Eiserne Adler* (The Iron Eagle), he survived being shot down in September by taking to his parachute. He was awarded the *Pour le Mérite* on 2 Nov. 1918; only two more aces received the award before the Armistice. The red-white-black band on the fuselage was Bäumer's personal marking.

Above & Below: Two views of Pfalz D.XIV 2800/18 at Adlershof during the *Typenprüfung* in July 1918. The D.XIV used the same basic structure as the D.XII, including the same fuselage and horizontal tail. The wings used the same ribs but the span was increased by one meter to give additional wing area to support the extra engine weight. The ailerons were enlarged to maintain roll rate with the longer wing span, and the vertical fin and rudder were enlarged to compensate for the increased engine power. The bump under the engine was necessary to provide room for the larger oil sump of the 200 hp Benz Bz.IV engine. A small gravity tank is mounted on the top wing just left of center. The aircraft wore standard Pfalz factory colors and markings for the time, including Pfalz silver-gray fuselage and fixed tail surfaces, white rudder, and five-color lozenge fabric on wings and elevators. An order for 50 was approved but only a handful were built.

Right: The Pfalz D.XV was developed from the D.XII as a 'wireless' design; elimination of the bracing wires eliminated significant drag and improved speed. Here a production Pfalz D.XV is at Adlershof during type testing. Production D.XVs had the 185 hp BMW D.IIIa engine, which used a pointed propeller spinner as seen here. The D.XV was arriving at *Armee Flugparks* the last week of the war, but was too late to see combat.

Left: A production Pfalz D.XV powered by the 185 hp BMW D.IIIa engine. Balances are fitted to all control surfaces to lighten control forces and make the airplane more maneuverable. With the D.XV Pfalz finally produced a first-rate fighter the equal of any operational fighter, and considered by the Germans to have flying qualities that matched the formidable Fokker D.VII. However, the D.XV was too late and Pfalz never overcame its reputation for second-string fighters.

Pfalz D.XV flown by Ernst Udet and later owned and flown by Max Holtzem for barnstorming in Argentina postwar. Holtzem had been a test pilot for Pfalz and later flew Pfalz D.IIIa fighters with *Jasta* 16b where he claimed two unconfirmed victories.

Left: Pfalz test pilot Gustav Bauer proudly stands before the prototype Pfalz D.X parasol monoplane in late 1918. Like the Pfalz Dr.I triplane and D.VII/D.VIII biplanes, it was powered by the 205 hp Siemens-Halske Sh.III counter-rotary engine. Although it did not officially compete in the Second Fighter Competition, it was one of the airplanes evaluated in July 1918 by front-line fighter pilots. Unlike the thick, cantilever Fokker wings, the Pfalz wing required extensive bracing, with the associated additional weight and drag, and the Pfalz D.X was not developed further.

Starting in mid-1917, Germany was consumed by the 'triplane craze' and triplane fighters were demanded by *Idflieg*, the German aircraft procurement authority. This was one of several serious missteps by *Idflieg* regarding fighters and wasted a large amount of precious engineering resources at a time when the new Allied fighters had regained technical superiority.

However, reality soon prevailed. Only one new triplane design was entered in the First Fighter Competition in January, and by the Third Fighter Competition in October monoplane fighters outnumbered biplanes in the competition in an attempt to achieve maximum performance.

Left: The Roland D.XVI powered by the production 205 hp Sh.IIIa rotary competed at the Third Fighter Competition as "an interesting type" despite not using the specified 185 hp BMW.IIIa engine. During most of 1917 *Idflieg* encouraged designers to try triplanes, but during 1918 monoplanes became increasingly popular for fighter designs due to their greater speed, and the last two Roland fighter designs to be built were monoplanes.

Above: During the Third Fighter Competition the Roland D.XVII fighter competed against the Fokker V29 parasol monoplane derivative of the Fokker D.VII; both fighters were powered by the 185 hp BMW.IIIa. The Fokker V29 wing design was superior structurally and aerodynamically to the Roland wing and gave better flying qualities with greater strength. Moreover, by late 1918 the competitive situation between German fighter manufacturers was much more difficult than just competing against Fokker's excellent designs. Little-known Kondor designed an excellent new parasol monoplane, their E.III & E.IIIa, with strong wings and good flying qualities. Veteran SSW and relative newcomer Daimler also delivered new parasol monoplane fighters of exceptional performance, with Daimler's V-8 finally reaching production and powering its entry, although too late for combat.

Right & Below Right: The Kondor E.IIIa was powered by the 160 hp Goebel Goe.III rotary; the E.IIIa had a spinner and full cowl.

Below: The Kondor E.III was powered by the 160 hp Oberursel Ur.III rotary. The E.III had a cutaway cowl and no spinner was fitted. No armament is fitted. This E.III competed in the Third Fighter Competition. High-power German rotaries were now becoming available.

Left & Below Left: The Austrian aviation industry closely followed developments of its German ally, and the trend toward monoplane fighters was soon noted. This Aviatik (Berg) 30.40 parasol monoplane fighter prototype was powered by a 150 hp Le Rhône(St) engine. This Steyr engine was based on a French design. Derived from the earlier 30.27 biplane prototype, it was the lightest fighter built in Austria-Hungary. Intended as an interceptor, it was fairly fast and demonstrated excellent climb during testing during July and August 1918, but proposed production could not be sanctioned until sufficient supply of the required Voltol lubricant could be assured.

Left: The Nieuport 31 monoplane fighter was the final Nieuport wartime design. Powered by a 170 hp Le Rhône rotary, it was planned to carry the typical two synchronized guns. The wheels were covered by a large auxiliary airfoil. No further development was undertaken, perhaps due to the success of the more conventional Nieuport 29 biplane and the declining interest in rotary engines.

Above: The SSW D.VI was not completed until after the armistice; by then the 'E' category for monoplane fighters was no longer used. Powered by the Sh.IIIa high-altitude counter-rotary engine then giving 205 hp, it had a top speed of 220 kmh (137 mph) coupled with excellent maneuverability and climb and a ceiling of 8,000 m. The under-fuselage fuel tank was jettisonable in case of fire. The SSW D.VI was the natural production successor to the SSW D.IV.

SSW D.VI Specifications

Engine:	205 hp Siemens-Halske Sh.IIIa	
Wing:	Span	9.37 m
	Area	12.46 m^2
General:	Length	6.50 m
	Height	2.700 m
	Empty Weight	540 kg
	Loaded Weight	710 kg
Maximum Speed:		220 km/h
Climb:	6000m	16 min
	7000m	22 min
Ceiling:		8000m

Daimler L11 Specifications

Engine:	185 hp Mercedes D.IIIbm Geared V-8	
Wing:	Span Upper	7.65 m
	Area	17.7 m^2
General:	Length	6.23 m
	Height	2.555 m
	Empty Weight	440 kg
	Loaded Weight	640 kg
Max Speed:		240 km/h
Climb:	2000 m	4.5 min.
	3000 m	7.7 min.
	4000 m	11.0 min.
	5000 m	13 min.

Above: The Daimler L11 fighter prototype was powered by Daimler's Mercedes D.IIIbm geared V-8 engine; here it is seen in its final configuration with vane-balanced ailerons. It had a top speed of 240 km/h, fast for its time.

Above: William Barker, VC, with his favorite Sopwith Camel B6313 in which he was credited with 46 of his 50 confirmed victories. Barker flew this Camel on both the Western and Italian Fronts and scored most of his victories in Italy.

William Barker was the most decorated Canadian in military service – a status he retains today. Despite achieving 50 aerial victories, most of Barker's decorations were awarded for his leadership in the air. He was highly regarded by the men who flew with him because they felt safe as part of his flight. Barker was not a lone-wolf looking to increase his victory score for self-aggrandizement. Unfortunately, the action for which he was awarded the Victoria Cross fits this stereotype of the Great War fighter pilot and has overshadowed Barker's prior achievements.

Above & Right: Barker's Sopwith Snipe, E8102, after his epic combat with numerous Fokker D.VII fighters on October 27, 1918 for which he was awarded the Victoria Cross. The remains of this fighter are preserved in Canada.

The official version of events as published in the London *Gazette* on 30 November 1918 reads in part:

"On the morning of the 27 October 1918, this officer observed an enemy two-seater over the Foret de Mormal. He (Barker) attacked this machine and after a short burst it broke up in the air. At the same time a Fokker biplane attacked him, and he was wounded in the right thigh, but managed, despite this, to shoot down the enemy aeroplane in flames. He then found himself in the middle of a large formation of Fokkers who attacked him from all directions, and was again severely wounded in the left thigh, but succeeded in driving down two of the enemy in a spin. He lost consciousness after that, and his machine fell out of control. On recovery, he found himself being again attacked heavily by a large formation, and singling out one machine he deliberately charged and drove it down in flames. During this fight his left elbow was shattered and he again fainted, and on regaining consciousness he found himself still being attacked, but notwithstanding that he was now severely wounded in both legs and his left arm shattered, he dived on the nearest machine and shot it down in flames. Being greatly exhausted, he dived out of the fight to regain our lines, but was met by another formation, which attacked and endeavored to cut him off, but after a hard fight he succeeded in breaking up this formation and reached our lines, where he crashed on landing. This combat, in which Major Barker destroyed four enemy machines (three of them in flames), brought his total successes to fifty enemy machines destroyed, and is a notable example of the exceptional bravery and disregard of danger which this very gallant officer has always displayed throughout his distinguished career."

Barker underwent numerous surgeries and never completely recovered from the serious wounds he received that day. Regardless, he was back in the air as soon as possible, and was very active in postwar aviation in his native Canada.

Above: The little-known Aviatik D.VII fighter was apparently placed in production but never reached the front. The Inter-Allied Control Commission found 50 production aircraft in clandestine storage after the war. Its 195 hp Benz V-8 was finally in production – too late for the war.

Right: The Spad 22 used a Spad 17 fuselage and engine coupled with a swept-back upper wing and a swept-forward lower wing with the chord narrower at the root than at the tip. This strange design indicated Spad was struggling to improve a basic airframe that had reached its development potential.

Below: The Rumpler D.I was one of the two winners of the Third Fighter Competition, but despite prolonged development was still not ready for full production nor operational service. Rumpler, justly famous for exceptional reconnaissance airplanes, exceeded the state of the art trying to develop the D.I and failed. By the Armistice the SSW D.VI and Daimler L.11 exceeded the D.I's performance without its structural and handling problems. Ernst Udet, a 62-victory ace (left), is shown in conversation with Edmund Rumpler in this postwar photograph.

Two-seat fighters like the Bristol F.2B, Halberstadt CL.II and CL.IV, and the Hannover CL.II/III/IIIa were very successful in combat, with their gunners providing a valuable 'sting in the tail' to defend against attack from the rear and also having the ability to attack aircraft off to the sides of the line of flight. With speed and maneuverability approaching or matching that of single-seaters, which could only fire forward and had no defense to the rear, these high-performance two-seat fighters were very effective in air-to-air combat and well-suited to ground attack. In fact, these two-seat fighters were so effective that Britain and France planned to field about twice as many two-seat fighters as single-seaters in 1919! Had this event actually occurred, it may have given an entirely different flavor to the mythos of the fighter pilot.

Germany had reached similar conclusions, particularly with regard to seaplane fighters; single-seat seaplane fighter development was essentially abandoned by the German Navy by the end of 1917. Germany also intended to field more two-seat fighters with the *Luftstreitkräfte* in 1919, although not in as high a proportion as the Allies were planning. To that end, German two-seat fighter development continued aggressively, and the new two-seat fighters were designed to withstand the same maneuvering loads (Gs) as single-seat fighters. In addition to follow-on developments of the well-known Halberstadt and Hannover two-seat fighters, the all-metal Junkers CL.I was just going into service at the Armistice and a seaplane version was under test, as was an all-metal two-seat seaplane fighter from the Zeppelin company.

Left: The Hanriot-Dupont HD.3 C2 two-seat fighter resembles the Salmson 2 A2 reconnaissance aircraft much more than the earlier Hanriot HD.1 fighter, probably because Hanriot had been building the Salmson under license and the HD.3 used basically the same engine as the Salmson, a 260 hp Salmson radial. The HD.3 had two fixed and two flexible guns and a very good performance. Eighteen were in service in October 1918, 15 of those with HD 174, but the Armistice resulted in further production being limited to 75 aircraft in total.

Left: The Société d'Études Aéronautiques (SEA) 4 two-seat fighter powered by a 370 hp Lorraine-Dietrich engine was ordered in great quantity in Spring 1918 but the first production aircraft was not delivered until Armistice Day. Only 115 of the order for 1,000 were built, the rest being cancelled. The aircraft built were used as long-range reconnaissance aircraft postwar.

This Page: Just as the BMW engine gave German single-seat fighters better performance, it was planned to use the BMW in the next CL-types, making them fifth generation fighters. The Hannover CL.V was tried with both biplane tail (top) and monoplane tail (above); the monoplane tail was placed in production. The Halberstadt CLS.I (right) was a smaller, faster, more maneuverable development of the CL.IV.

Above: The Spad 11 was developed from the single-seat Spad 7 and intended as a two-seat fighter. Unlike its single-seat predecessor, it had poor maneuverability and handling qualities and could not perform as a fighter. Its speed was fairly good so it was widely used as a reconnaissance aircraft.

Above: The Spad 16 was simply a Spad 11 with different engine, indicated by the protruding cylinder heads; neither performance nor handling qualities were improved but the Spad 16 served in numbers as a reconnaissance plane. American General 'Billy' Mitchell used one for his personal transportation and is here standing in front of his personal aircraft with his pilot. This aircraft (or a similar aircraft) is preserved in the National Air & Space Museum.

Above & Below: The unlovely Morane-Saulnier AN series of two-seat fighters marked a complete departure for the company, which had previously created many elegant designs. The design requirement was for a fighter/reconnaissance aircraft, and armament was a synchronized gun for the pilot and two flexible Lewis for the observer. Despite a slow rate of climb, the AN was ordered into production, but the reliability of its Bugatti engine apparently was suspect and other engines were fitted. The ANB (top left) was powered by the 450 hp Bugatti. The ANL (above right) was powered by the 400 hp Liberty, and the ANS (below left) was powered by a 530 hp two-row Salmson radial. The ANR (below right) was powered by a 450 hp Renault V-12. Only one prototype of each variant was built; the Armistice reduced interest in the AN and the competing SEA 4 was already in production.

By 1918 no nation was more interested in two-seat fighters than France, and these pages illustrate some of the many French two-seat fighter designs used or tested in 1918.

Right: The Spad 20 two-seat fighter was far more compact than the earlier Spad 11 and Spad 16 that were intended as two-seat fighters but were relegated to reconnaissance duties due to poor maneuverability and handling. Powered by a 300 hp Hispano-Suiza, it was fast and maneuverable, just the aircraft Spad had been trying to produce. The war ended before production aircraft could be delivered, and only 95 were built.

Right: The Borel 3000 was another late-war two-seat fighter design powered by a 300 hp Hispano-Suiza. Two guns were synchronized for the pilot and the observer had two flexible guns. The Borel 3000 had excellent performance but by the time it was tested the Spad 20, which used the same engine, had already been chosen for production to fill the same role.

Right: The B.A.J. C2 two-seat fighter was another design powered by a 300 hp Hispano-Suiza. The pilot had a single synchronized gun and the observer had two guns. It did not fly until after Armistice Day, and only two aircraft were built.

Sixth Generation Fighters

Sixth Generation fighters were those with all-metal structures. The list of all-metal WWI fighters is very short and includes only those aircraft designed by two companies, Junkers and Zeppelin.

Professor Hugo Junkers became interested in aviation in 1907 and began to explore the 'thick-wing' concept in 1912. By late 1914 he had decided that an all-metal aircraft was practical and in February 1915 he approached the War Ministry for money to build an all-metal airplane. Initially he was turned down, but further research and experimentation in metal forming convinced the War Ministry to provide the funds in July 1915.

The first aircraft, Junkers internal factory designation the J 1, was an experimental two-seater built of thin sheets of steel because techniques for forming duralumin were not yet adequate for aircraft construction. The J 1 all-metal monoplane made its first hop during ground tests on December 12 and its first full flight on January 18, 1916. A streamlined, under-fuselage radiator was fitted. The J 1 was fast and robust but understandably over-weight.

Sixth Generation Fighters	At Front
Junkers CL.I	~11/18– Armistice
Junkers CLS.I	No Service
Junkers D.I	~11/18–Armistice
Zeppelin-Lindau CS.I	No Service
Zeppelin D.I	No Service

Above: The Junkers J 1 (an internal company designation) was the first Junkers aircraft to fly. It was a two-seat test-bed for Junker's all-metal structural technology, and the Junkers J 2/E.II fighter was developed from this aircraft.

Junkers J 2/E.II

The J 1 was sufficiently successful that *Idflieg* ordered the construction of six similar all-metal fighters as the J 2, military designation E.II, with serials E.250–256/16. Like the J 1, the E.II fighters were made of steel. Again a streamlined, under-fuselage radiator was fitted. The wing was mounted low on the fuselage for the best lift/drag ratio; furthermore, in event of a crash the low-mounted wing would give better protection to the pilot. A rollover structure was mounted directly behind the

Above: The Junkers J 2 (an internal company designation) was given the military designation Junkers E.II. Six examples were built; E.251/16 is shown here. The single synchronized LMG 098/15 machine gun was conveniently located.

cockpit to protect the pilot if the aircraft flipped over. The first flight of a J 2/E.II was on June 11, 1916.

The first aircraft built was powered by a 120 hp Mercedes D.II engine, but the five subsequent aircraft used the more powerful 160 hp Mercedes D.III engine. A single LMG 08/15 machine gun was installed. A number of changes were made to the prototypes that resulted in improved flying qualities, and E.253/16 was fitted with a wing of increased span and area in an attempt to improve performance. Streamlining made the E.II fast but the airplanes were much too heavy due to their steel construction, resulting in poor climb, ceiling, and maneuverability, so they were not suited for combat or production.

Junkers J 7

After the steel J 2 the obvious next step for Junkers was to build his aircraft out of duralumin to reduce weight. Work on techniques for building an airplane out of duralumin had continued in parallel with construction of the E.II prototypes and the J 7 single-seat fighter prototype was built from duralumin, with work starting in the late spring of 1917 in parallel with the J 8 two-seat fighter prototype.

Idflieg had banned belly and ear radiators on combat aircraft on November 10, 1916 because the coolant drained out much more quickly if there was a leak, so the J 7 was designed for a nose radiator. Otherwise the J 7 was similar in configuration to the earlier J 2 but was smaller and much lighter. The rollover structure was now enclosed in a streamlined headrest. A 160 hp Mercedes D.III engine was fitted.

Initially, the radiator was mounted above the engine to expedite flight-testing. At first the J 7 had rotating wingtips instead of conventional ailerons, but in flight these immediately demonstrated they were over-balanced. A new wing with conventional ailerons was soon installed for further flight trials. Even with the drag-producing radiator mounted above the engine the J 7 was fast. When flown in comparison to a standard Albatros D.III on October 22, 1917, by naval aces *Lt.* Theo Osterkamp and *Lt.* Gotthard Sachsenberg, the J 7 easily out-climbed the Albatros and was also significantly faster.

On December 1, 1917, the J 7 was tested with a nose radiator, new ailerons, and revised rudder. These changes greatly improved flying qualities while also making the aircraft faster. Fitted with yet another new wing with aerodynamically-balanced ailerons, the J 7 was entered in the First Fighter Competition in January 1918 and impressed the pilots with its speed and climb. Development continued and resulted in the production Junkers D.I fighter, the first all-metal fighter to go into series production and combat.

Above: The Junkers J 7 prototype as originally built with rotating wing-tip ailerons. The long exhaust pipe directed the exhaust away from the pilot, and a sturdy roll-over structure was enclosed by the streamlined headrest. The radiator was mounted over the engine to expedite flight testing while investigation into the ultimate configuration of the nose radiator mounting continued.

Above: The Junkers J 7 prototype photographed on October 12, 1917, after being fitted with a new wing using conventional ailerons. The long exhaust pipe and test equipment are still fitted, and the original radiator was still mounted over the 160 hp Mercedes D.III. A neat spinner is fitted over the propeller hub.

Above: The Junkers J 7 prototype in its third configuration photographed January 1918 while being readied for the First Fighter Competition. It now has yet another new wing with aerodynamically-balanced ailerons, and a nose-mounted radiator. The spinner and the long exhaust pipe are both gone and it has a new rudder.

Above: The Junkers J 7 prototype photographed on February 2, 1918, in its fourth configuration after being fitted with yet another new wing of longer span. The ailerons are also longer and no longer have aerodynamic balances. Test equipment are still fitted. The test pilot is Arved Schmidt.

Above & Below: The Junkers J 7 prototype in its fifth configuration photographed March 23, 1918 with larger radiator for summer operation. The aircraft was accepted by *Idflieg* for demonstrations and training the week of March 25, 1918.

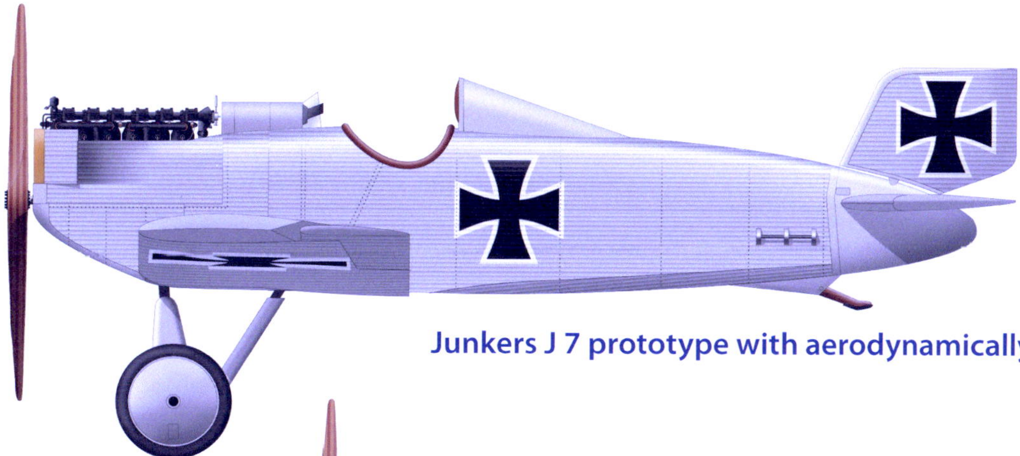

Junkers J 7 prototype with aerodynamically balanced ailerons

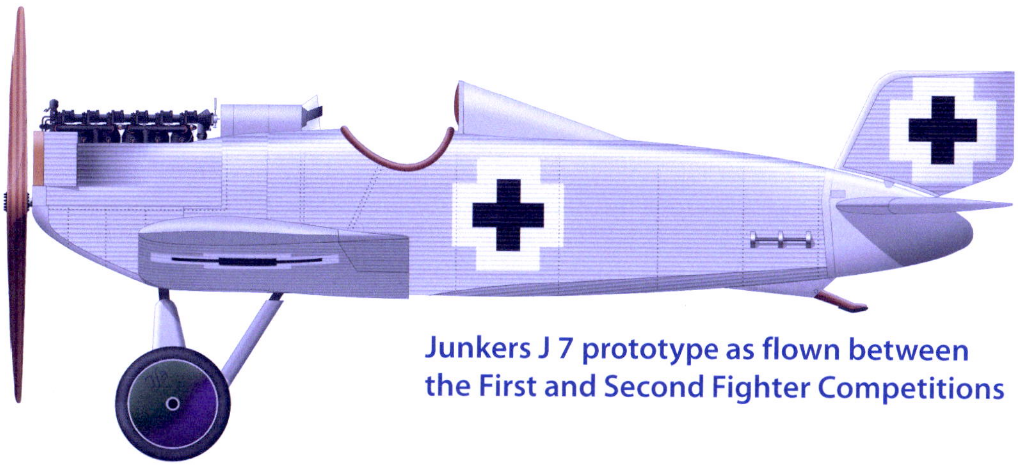

Junkers J 7 prototype as flown between the First and Second Fighter Competitions

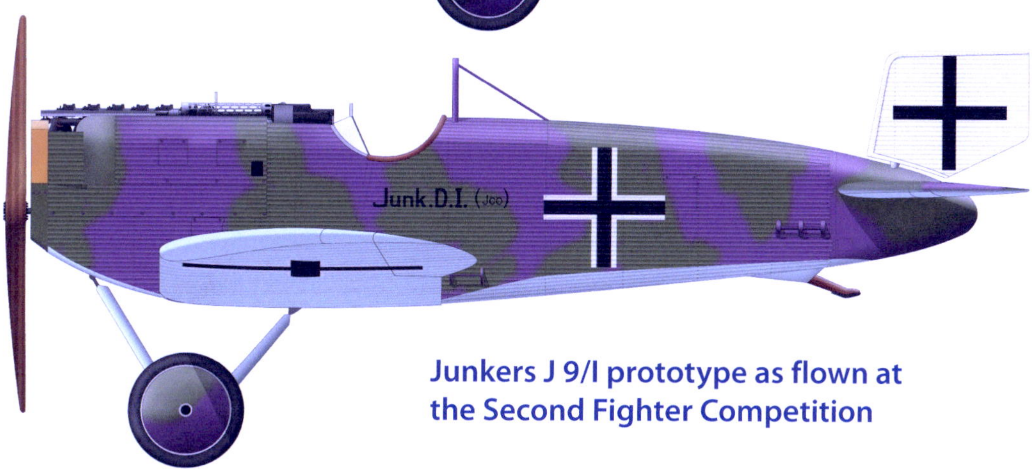

Junkers J 9/I prototype as flown at the Second Fighter Competition

Junkers D.I 5180/18 with longer fuselage and wings. An aircraft like this without camouflage was flown at the Third Fighter Competition

Junkers J 9/D.I

After the First Fighter Competition the J 7 was further modified and tested, and was then purchased as a trainer and demonstration aircraft. It was again demonstrated at the Second Fighter Competition along with the revised J 9, the D.I production prototype. Despite its good performance and the quality of its engineering and construction, some prejudice remained against is low-wing configuration and all-metal construction.

Junkers began work on the J 9, the prototype for the production D.I, in February 1918. The focus of the two prototypes ordered was revising the basic J 7 airframe to facilitate mass production. On May 8 Junkers received an open contract for 100 all-metal aircraft, the first 20 of which were D.I fighters. The first flight of a prototype D.I fighter, referred to as the J 9/I, is thought to be on May 12.

This aircraft was readied for the Second Fighter Competition with a 180 hp Mercedes D.IIIaü high-compression engine that gave improved performance at high altitude. At the competition the J 9/I out-climbed all other Mercedes-powered fighters except the Rumpler D.I. In fact, the J 9/I was lighter than the competing Fokker V21 and V23 monoplanes made of wood.

The second D.I prototype, the J 9/II, was completed in June. Powered by the experimental 195 hp Benz Bz.IIIbo un-geared V-8, it was intended to compete at the Second Fighter Competition but the engine was too unreliable.

Junkers D.I Specifications

Engine:	185 hp BMW.IIIa	
Wing:	Span	9.00 m
	Area	14.8 m^2
General:	Length	6.70 m
	Length (late version)	7.25 m
	Height	2.60 m
	Empty Weight	654 kg
	Loaded Weight	834 kg
Maximum Speed:		225 km/h
Climb:	5000 m	24.0 min.

Despite demonstrating superior performance and flying qualities, the front-line pilots considered the D.I's low-wing monoplane configuration a 'total failure'! According to the Austro-Hungarian liaison officer at the trials, the reason for that harsh judgment was the low-mounted wing greatly obscured the downward field of view from the cockpit. By this time German fighter pilots preferred 'dive and zoom' tactics to avoid getting entangled with large numbers of enemy fighters, and that put a premium on good downward visibility, which was limited in the Junkers fighter. By WW2 the low-wing monoplane configuration was universal for fighters, rendering the pilots' judgment of 'total failure' ridiculous in retrospect.

Development of the Junkers D.I continued between the Second and Third Fighter Competitions due to *Idflieg's* support, and the type-test aircraft,

Below: The Junkers J 9/I production prototype of the D.I was photographed July 8, 1918 at the Junkers air field at Dessau.

Above & Below: The Junkers J 9/II prototype was powered by the experimental 195 hp Benz Bz.IIIbo un-geared V-8 engine. Intended to compete at the Second Fighter Competition, the unreliable engine prevented its appearance. The under-fuselage radiator was modern for its time.

5180/18, was approved for front-line service on 9 September 1918 after final type testing.

For the Third Fighter Competition Junkers entered a D.I fitted with BMW engine, longer fuselage, and increased wingspan. The enlarged dimensions reduced maneuverability and performance. However, the Junkers D.I still bested all but one of the competitors in rate of climb and was faster than the Fokker V36, although the Fokker V29 was faster above 4,000 meters.

Now joined by the all-metal Zeppelin D.I biplane fighter using advanced, stressed-skin, semi-monocoque construction, the Junkers D.I was no longer the world's only all-metal fighter, but it was already in production as the world's first all-metal *production* fighter. Several production aircraft reached Belgium before the end of the war, but nothing is known of their combat use, if any. However, one airframe had been struck 'by several bursts of machine gun bullets,' an indication they had experienced combat.

Postwar the D.I was used very successfully in combat in severe winter weather and stood up to the elements far better than wooden aircraft. According to *Lt*. Sachsenberg who flew them, "The Junkers have proven themselves beyond all expectation."

Above & Below: The Junkers CL.I was the two-seat fighter development of the single-seat D.I. Larger than the D.I, it was powered by a 170 hp Mercedes D.IIIa engine and was armed with two fixed, synchronized guns for the pilot and a flexible gun for the observer. Developed from the very similar J 8 prototype, the production aircraft, first ordered in March 1918, was known as the J 10 to the factory. The CL.I went into production in the second half of 1918 to supplement the existing Halberstadt and Hannover CL-types, and 44 had been delivered through March 1919. The CL.I, along with the Junkers D.I, was used by the *Geschwader* Sachsenberg during the postwar fighting against the Bolshviks in the Baltic. Both the D.I and the CL.I were very robust aircraft that served successfully in severe climatic conditions. The CLS.I was a floatplane derivative of the CL.I.

Zeppelin D.I

The Zeppelin company is normally thought of in conjunction with airships. In fact, such was the prominence of the Zeppelin company's rigid airships that large, rigid airships are often called 'Zeppelins' regardless of manufacturer, However, Zeppelin also built Giant bombers at its Staaken facility and designed and built giant, all-metal flying boats at its Lindau facility. These all-metal flying boats were designed by *Dipl.-Ing.* Claude Dornier, later to become well-known as founder of the Dornier aircraft company before WW2.

Dornier was also the designer of the Zeppelin D.I all-metal fighter, sometimes known as the Dornier D.I after its designer. The Junkers company built the world's first all-metal production aircraft, the Junkers J.I, and also the world's first all-metal production fighter, the Junkers D.I. But the Zeppelin D.I was the world's first *stressed-skin* all-metal fighter, or fighter in which the metal skin served as an integral, load-bearing part of the structure. The Zeppelin D.I was the most structurally-advanced airplane to fly in WWI, and its structure was similar to that of modern all-metal aircraft.

The proposal for the Zeppelin D.I was approved by *Idflieg* on 28 February 1918. Six prototypes were ordered, D.1750–1755/18, three to be powered by the 170 hp Mercedes D.IIIa and three by the experimental 195 hp Benz Bz.IIIb V-8 then in development. However, the V-8 was never installed in a Zeppelin D.I, the BMW.IIIa being used instead.

Zeppelin D.I Specifications		
Engine:	170 hp Mercedes D.IIIa	
Wing:	Span	7.80 m
	Area	16.66 m^2
General:	Length	6.40 m
	Height	2.60 m
	Empty Weight	725 kg
	Loaded Weight	883 kg
Climb:	5000m	25 min

One June 4, 1918 the first prototype, D.1752/18, was taken for its first flight by test pilot *Vzfw.* Heinz Ruppert. The second D.I prototype, D.1751/18, competed at the Second Fighter Competition. It was flown by a number of pilots, then *Hptm.* Wilhelm Reinhard, Commander of *JG*I since the death of the Red Baron and a 20-victory ace nominated for the *Pour le Mérite*, took it up on 3 July. Reinhard put the prototype D.I through its paces, and while pulling out of a steep dive the top wing tore off and Reinhard was killed in the ensuing crash. This tragedy should not have happened; the D.I had not yet passed its official static load tests and regulations prohibited front-line pilots from flying aircraft until all load tests were passed.

Regardless of the tragedy, which inspired the fictional scene in the book and movie *The Blue Max*,

Above: The all-metal Zeppelin D.I 1752/18, the first protoype to fly, is seen here. Zeppelin D.I 1751/18, the second prototype to fly, was entered in the Second Fighter Competition where, during an evaluation flight on July 3, 1918, the upper wing tore off. *Hptm.* Wilhelm Reinhard, who had succeeded Manfred von Richthofen as the commander of *JGI*, was killed in the subsequent crash. *Oblt.* Hermann Göring had flown the fighter immediately prior to Reinhard.

the combat pilots who had flown the Zeppelin D.I all agreed that it was on average superior to the other fighters powered by a Mercedes engine. The D.I was said to have flown tighter turns than the Fokker D.VII with about equal maneuverability. In addition to its maneuverability, the D.I had good speed, very good rate of climb, and excellent flying qualities.

The Zeppelin's obvious potential was such that development was continued. After the wings and their attachments had been reinforced, airframe D.1750/18 was tested to destruction from late September to 11 October and easily passed the load test requirement, which had been raised to 6.5 Gs from 5 Gs after Reinhard's fatal accident.

Above: The all-metal Zeppelin D.I 1752/18, the first protoype to fly, displays its clean, innovative lines.

Below: A Zeppelin D.I with 185 hp BMW.IIIa being tested by the U.S. Army at McCook Field in 1922.

Right: Zeppelin D.I 1751/18 (Ersatz) is seen here at the Third Fighter Competition. It was a replacement for the original D.I 1751/18 destroyed at the Second Fighter Competition. Powered by the 185 hp BMW.IIIa, a similar aircraft was evaluated by the US Air Service at McCook Field in 1922. The fuel tank was under the fuselage to minimize danger from fire, and reportedly was planned to be jettisonable.

Zeppelin D.I D.1751/18 (*Ersatz*), powered by a 185 hp BMW.IIIa engine that substantially improved the rate of climb, was built to replace the aircraft Reinhard crashed. *Idflieg* ordered a batch of 50 Zeppelin D.I fighters but production was halted in early 1919 and only a few were completed. Postwar the US Air Service purchased one and evaluated it as AS-68546. Similarly, the US Navy evaluated another D.I, D.1753/18. The final prototype was exhibited in the Dornier Museum in Friedrichshafen but was destroyed by Allied bombing during WW2.

Zeppelin D.I (Ersatz) Specifications

Engine:	185 hp BMW.IIIa	
Wing:	Span	7.80 m
	Area	16.66 m^2
General:	Length	6.40 m
	Height	2.60 m
	Empty Weight	725 kg
	Loaded Weight	883 kg
Maximum Speed:		200 km/h
Climb:	1000m	2.6 min
	5000m	13 min
	7600 m	51 min

Below: The all-metal Zeppelin D.I fighter had the most advanced structure of any WWI aircraft. Prototypes were powered by both the Mercedes D.IIIa and by the 185 hp BMW.IIIa. The surface of the wings aft of the box spar had fabric covering, as did the horizontal tail surfaces. This example was brought to the USA for testing after the war and is shown after re-assembly. Other than its biplane configuration, this was the preview of future propeller-driven fighters until the jet age; then early jets used similar structural technology. Postwar the Zeppelin D.I was often referred to as the 'Dornier D.I' in honor of its brilliant designer, *Dipl-Ing.* Claude Dornier, who later founded his own company, but Zeppelin D.I, the company name during the war (or Zeppelin-Lindau D.I after the division of the company), is what it was called when it was designed, built, and flown and therefore is the correct designation. The Zeppelin-Lindau CS.I is described and illustrated in the floatplane section.

Other Fronts

The Western Front was the critical front during WWI, and that was especially the case for aviation. Nevertheless, fighters fought in other places and the combat was just as dangerous.

Although many well-known fighter types were used over these other fronts, some fighters were unique to these areas remote from the main event. But pilots could be killed just as dead.

The Italian Front

Above: Italian soldiers walk near Albatros and Phönix fighters lined up on Bressanone airfield in a cold, foggy November day immediately after the end of the war. (Aeronautica Militare)

Before WWI Italy was loosely allied with Germany and the Austro-Hungarian Empire, but decided to stay neutral when war broke out. Considering her opportunities, on 23 May 1915 Italy declared war on the Austro-Hungarian Empire in hope of winning territorial gains, such as the city of Trieste. The Italian war death toll of 651,000 soldiers and 589,000 civilians, nearly 3.5% of Italy's population of 35.6 million, was a higher price than expected.

Italian industry was more developed than that of the Austro-Hungarian Empire, her primary adversary. Despite the usual combination of technical problems and managerial mistakes, Italy produced more than twice as many airplanes and three times as many aero engines as Austria, not least because Austria was plagued with similar technical and managerial problems. This far larger aviation production gave Italy a long-term air superiority advantage over Austria. Moreover, as an ally of France, Italy was able to import French airplanes and aero engines as well as produce them under license in Italy.

Although Austria was able to import a few German airplanes for her air service and built a number of German designs under license, Germany never had an adequate number of aero-engines and sold very few to Austria. Inadequate production of aero-engines thus hobbled Austria's *Luftfahrtruppen* throughout the war.

Only one original Italian fighter design, the Balilla, saw limited service, and only scored one confirmed victory. Produced for commercial, not military, reasons, the Balilla was a fast airplane but was not maneuverable enough for combat with other fighters and had not solved its technical problems before the Armistice. The French Hanriot HD.1, not used by France itself, which favored the SPAD, was the favorite fighter in Italian service.

The Austro-Hungarian aviation story is not as happy as that of Italy. Largely agricultural and riven by ethnic strife, the Austro-Hungarian Empire was less developed industrially and had fewer raw materials, which greatly hampered its airplane and aero-engine production.

Above: *Sergente* Giovanni Bartolomeo Arrigoni of 76ª *Squadriglia* warms up his Nieuport 11 on Borgnano airfield. The Nieuport 11 was the first effective fighter employed by the Italians. (Aeronautica Militare)

When fighter airplanes were needed, Austria did not have a satisfactory design available and bought some Fokker *Eindeckers* from Germany. The first fighter produced in Austria was the Brandenburg D.I, a German design built under license but not used by the Germans, which should have been a warning. While strong and fast, the D.I (also known as the KD, for *Kampf Doppeldecker* – battle biplane) had poor, perhaps even vicious, handling characteristics, and more pilots were killed in flying accidents than were downed in combat. The KD had another problem; due to the shortage of synchronizers, its gun was mounted above the upper wing, but in a streamlined container that prevented the pilot from clearing jams in flight.

Noting the problems of the KD while producing it under license, the Phönix company developed the Phönix D.I fighter starting with the KD as a basis. The wings and struts were redesigned and a more powerful engine was fitted. A number of prototypes were built and test flown before handling characteristics were acceptable, and the resulting Phönix D.I was a completely reliable, maneuverable fighter with two synchronized guns. Development continued with the D.II; better detail design reduced weight and improved maneuverability. The D.IIa, with more powerful engine and better performance, followed in production, which was limited by engine shortages. The improved D.III and D.IV were too late for combat.

The Austrian Aviatik company, with *Oberingenieur* Julius von Berg as the designer, produced the Aviatik (Berg) C.I two-seat reconnaissance airplane and the D.I single-seat fighter. Both offered good performance and the D.I in particular had good maneuverability and handling characteristics. Unfortunately, both were fragile designs plagued by structural failures, especially the fighter version, which was subjected to greater maneuvering stress in combat.

Interestingly, the best Austrian fighter to see combat was probably the Albatros D.III built under license. The major shortcomings of the KD lead Austrian authorities to move with uncommon speed to seek a suitable replacement at the time the Albatros fighters were proving themselves on the Western Front. The Austrian Oeffag company obtained a license from Albatros, and production of the Albatros D.II was well underway when the contract was signed for the airplanes on 4 December 1916! Only 16 D.II fighters were produced, production then switching to the Albatros D.III. Most of the D.II fighters were sent to the less demanding Russian Front, where it proved to have excellent climb and good handling characteristics,

Left: The Hanriot HD.1 was a French design used by Belgium and Italy but not France, which preferred the sturdy Spads. Powered by the same engine used in the Nieuport 17, the HD.1 had a superior wing design that was more robust and offered somewhat better performance and maneuverability than the Nieuport. Italian pilots liked the HD.1 and it became the most numerous Italian fighter.

but was considered too slow – the eternal fighter pilots' lament!

When Oeffag undertook Albatros D.III production they used a more powerful engine than the German original (185 hp compared to 160 hp) and, more importantly, they strengthened the wing and fuselage. The result was a fast, tough, maneuverable fighter that was able to use more powerful engines as they became available and that had no problems with wing failure as suffered by Albatros fighters, built in Germany. The Albatros D.III(Oef) was produced in quantity in several series, the final 253 series featuring a 225 hp engine reaching the Front in June 1918. *Flik* 63J considered the new fighter as "equal to all combat requirements, it is very much liked, only the squadron does not receive sufficient numbers". *Flik* 61J considered it "first class and superior to any fighter".

French and British Assistance

During Italy's great Caporetto defeat in late 1917, in which some German army and aviation units participated, Italy lost many airplanes and aircrew, and with them the air superiority she had

Above: The Balilla (Hunter) was the only fighter of Italian design to see combat. Like the SVA 5 from which it was developed, it was fast but lacked maneuverability. Purchased despite offering no superiority over the Spad 13 already in service, the Balilla scored only one confirmed victory during the war.

Left: Italian SVA 5 of the 87a *Squadriglia* on a reconnaissance mission over the Alps; three streamers fly from the tailplane. The SVA was the fastest airplane on the Italian front and was an exceptional single-seat reconnaissance aircraft. Not maneuverable enough to be very successful against fighters, it did down some two-seaters.

Above: This portrait of Francesco Baracca, the leading Italian ace, was taken on 10 May 1918. Baracca was now a Major and the commanding officer of the 91st Squadron. This SPAD 13, S.2445, was retired on 21 May, and he was flying a SPAD 7 when killed on 19 June 1918. Baracca's personal marking may seem familiar; it was later adopted by Enzo Ferrari for his racing cars as the suggestion of Baracca's mother. The horse was black and Ferrari added a canary yellow background to his version of the insignia because it is Modena's color. (Via Paolo Varriale)

Above & Above Right: Italy claimed that Major Baracca was killed by ground fire. However, like many other events in the air war, there is more than one version of the death of Major Baracca. He was also claimed by the crew of *Oblt*. Arnold Barwig and *Zugsführer* Max Kauer in Phönix C.I 121.17 pictured above. Although a reconnaissance pilot, Kauer achieved a total of four victories in this aircraft.

Which version of the story is true may never be known for certain. In any case, a well-flown two-seater was a tough opponent for any fighter pilot, and the Phönix C.I was the best Austro-Hungarian reconnaissance airplane. Phönix C.I 121.16 above shows the unique tail design that gave observers in this type such a wide field of fire. The Phönix C.I was also fast, tough, and maneuverable.

Phönix-built D.I 28.30 as flown by *Hptm*. Raoul Stojsavljevic, CO of *Flik* 16. The gun mounting on this pioneering *Fotoeinsitzer* was non-standard.

Berg D.I 138.99 flown at *Flik* 42J by *Off.Stv*. Friedrich Hefty. On 22 August 1918 Hefty, who scored 5 victories, became the first Austro-Hungarian pilot to parachute to safety when he bailed out of his burning Albatros D.III(Oef) 253.71.

Albatros D.III(Oef) 253.116 flown by 10-victory ace Oblt. Freidrich Navratil, CO of *Flik* 3J.

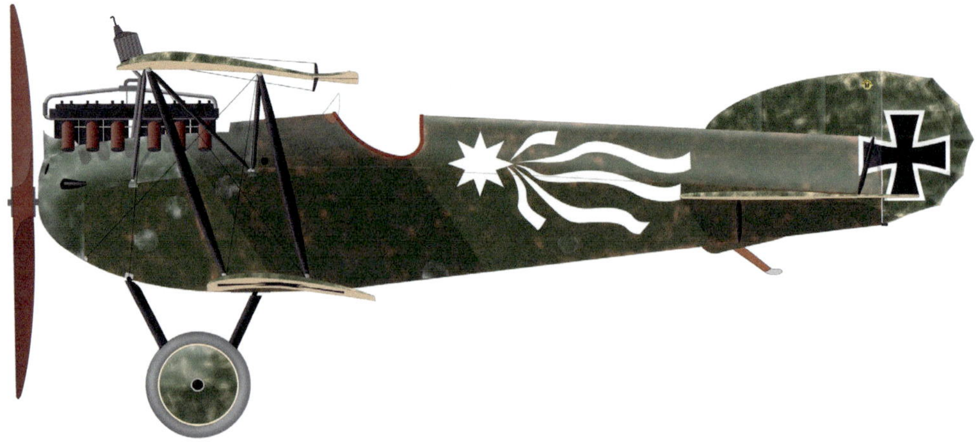

Phönix D.IIf flown by Oblt. Hans Leiner of *Flik* 55J from Pergine Aerodrome in July 1918. The comet was Leiner's personal marking.

Left: This star bedecked, Phönix-built Brandenburg D.I was 28.10. Allocated to *Flik* 41J on 24 February 1917 while the unit was forming up at Strasshof, it was at the Front in April. Flown by Godwin Brumowski, later to become the leading Austro-Hungarian air ace, as his first fighter, it was also flown by Samek and Richter amongst others. It was later given the modified tail surfaces. Note the Phönix decal on the rudder. *(via AHT AL0579-042)*

The distinctive struts of the Brandenburg D.I make the type easy to identify; postwar the D.I was given the nickname 'star-strutter.' Although strong and fast, the D.I (also known as the KD, for *Kampf Doppeldecker* – battle biplane) had poor, perhaps even vicious, handling characteristics, and more pilots were killed in flying accidents than were downed in combat. Very few D.I fighters had the benefit of a synchronized machine gun, with the result that most carried a gun in a canister firing over the upper wing.

Above & Right: Phönix-built Brandenburg D.I 28.30 was an armed, single-seat reconnaissance fighter; note the camera pod below the cockpit. *Hptm*. Raoul Stojsavljevic, CO of *Flik* 16, was the pilot of this pioneering *Fotoeinsitzer*.

Above: The Phönix D.I (328.25 shown) was developed from the Brandenburg D.I. It was more powerful, fitted with two synchronized guns, and most important had improved maneuverability and handling qualities.

Below: The Phönix D.I was also used by the Austro-Hungarian Navy as an interceptor to defend naval air stations.

Left: Many batches of the Aviatik (Berg) D.I fighter served with the Austro-Hungarian *Luftfahrtruppen*. The Berg D.I was more maneuverable than the Phönix D.I and had a better climb rate, but was also more lightly built and more fragile. Wing spar failure in combat is thought to have resulted in the loss of *Oblt.* Frank Linke-Crawford, fourth-ranked ace with 27 victories, while flying this D.I.

Above: *Oblt.* Frank Linke-Crawford, commanding officer of *Flik* 60/J with 27 victories, on Feltre airfield summer of 1918.

Above Right & Right: The leading Austro-Hungarian ace, *Oblt.* Godwin Brumowski, survived the war with 35 victories. He is seen above right in a formal portrait and right with his personal insignia on one of his Albatros fighters.

Below: Brumowski and Linke joke in front of their fighters on Torresella airfield in December 1917. (Rudi Höfling)

established over the Front. During the disastrous retreat the Italian air force gave its best, recklessly attacking enemy ground troops to stop the advance despite heavy losses to ground fire. Many airfields were overrun by enemy troops and many airplanes were lost on the ground as a result. Italy requested Allied assistance and some French and British air units were quickly sent along with many French-built Hanriot and SPAD fighters. The British and French help stabilized the situation until Italy could re-group and re-equip her air service, and by June 1918 the Italians had regained air superiority, which they held until the Armistice.

The final Italian offensive started on 24 October 1918 and resulted in the Austro-Hungarian collapse. During the war both sides fought bravely and skillfully, but the greater industrial potential of Italy generally gave her air superiority, and in the end Italy prevailed.

Above: Pleasing photo of Francesco Baracca, the leading Italian ace of the war.

Below: Spad 7 perhaps flown by Baracca that is now displayed in the Baracca Museum.

Above: Prototype Austro-Hungarian WKF 80.05 triplane fighter. Its biplane development offered better performance and was placed into production, but too late for combat.

Above: Reproduction Sopwith Camel in flight; this aircraft is in the markings of Canadian ace William Barker, who flew this Camel in Italy and scored a number of victories there.

Above & Right: Aviatik D.II 39.01 was the first of 19 fighters of this type ordered for high-speed investigation, including brief front-line evaluation in July 1918. Appearing at Aspern for testing in August 1917, the D.II was the fastest Austro-Hungarian fighter of the war. However, it was not structurally robust, aileron effectiveness was sluggish, and climb rate was modest. Due to these limitations, no further production was planned.

Spad 13 of *Majore* Francesco Baracca, with 34 victories Italy's leading ace. After the war his rampant stallion personal insignia was adopted by Enzo Ferrari for his racing cars.

Hanriot HD.1 of Ten. Silvio Scaroni, with 26 victories Italy's second-ranking ace.

Nieuport 27 N.19750 of *Sgt*. Marziale Cerutti, 79a *Squadriglia*, Italian Air Service. MIR stands for "*Marziale Imperatore Romano*" (Marziale the Roman Emperor). Cerutti scored 17 victories and was tied for honors as Italy's sixth-ranking ace.

Above: *Capt*. Mario Zobobi, CO of 81a *Squadriglia*, stands beside his colorful Hanriot HD.1.

Above: Albatros D.III(Oef) 153.10, *Flik* 41J, of Godwin Brumowski, the leading Austro-Hungarian ace; the woman in the cockpit is unknown – but was not his wife! Pilots considered the final 253 series Albatros fighters with 225 hp engine the equal of any Allied fighter.

Above: Reproduction Bristol Fighter from The Vintage Aviator Collection in New Zealand. The Camel and Bristol F2b were two of the key types Britain sent to assist Italy.

Above: Three white stripes identify Phönix D.IIa 422.14 flown by *Zugsführer* Alexander Kasza. On this plane he obtained one victory, shooting down British Bristol F2b C4700 on 9th June 1918. (Gregory VanWyngarden)

Bristol Scout 8996 flown by FSL. G.T. Bysshe of No.2 Wing, RNAS at Imbros. On 17 February 1917 Bysshe was shot down in this aircraft by *Off.Stv*. Emil Meinecke; 8996 was then repaired and Meinecke flew it Turkish markings!

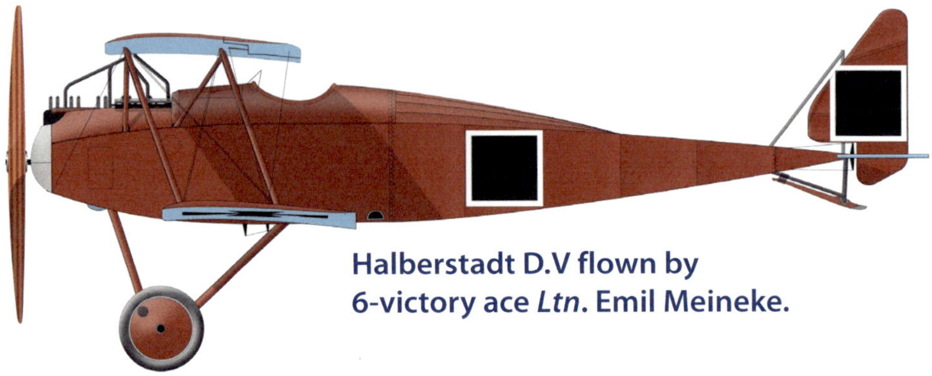

Halberstadt D.V flown by 6-victory ace *Ltn*. Emil Meineke.

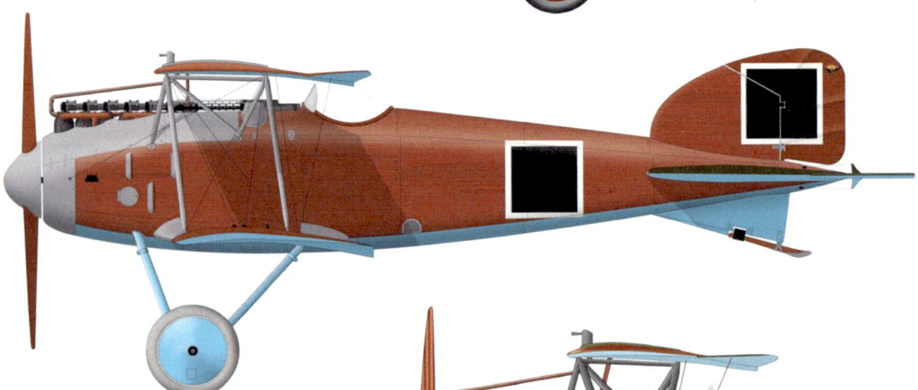

Albatros D.II flown by 6-victory ace *Ltn*. Emil Meineke.

Albatros D.III AKD3, Turkish Air Service, 1918.

Pfalz D.IIIa 1359/18 of the batch of 20 supplied to Turkey.

SE.5a of No.111 Sqdn., RAF, Palestine, 1918.

Martinsyde G.102 Elephant A1593 of No.67(A) Sqdn., RFC, Middle East, May–August 1917. A1593 was presentation aircraft *MOUNT LOFTY. SOUTH AUSTRALIA*.

Bristol F.2B of No.1 Sqdn., AFC, Palestine, 1918.

Bristol M.1B A5140 of No.111 Sqdn, RFC, Deir-el-Belah Aerodrome, late 1917.

Bristol M.1C of No.72 Sqdn, RAF, Palestine, 1918.

The Eastern Front

Above: With 20 victories, Maj. Aleksandr Kosakov was the leading Russian ace of the war. Flying with the British against the Bolsheviks and depressed by Bolshevik successes during the Russian Civil War, he deliberately dove his fighter into the ground.

The combat environment on the Eastern Front was substantially different than that on the Western Front. The large distances and lower population density permitted a war of movement in the East, although transportation capability via roads and railroads was poor compared to the West. Additionally, neither Russia nor Austria-Hungary were highly industrialized empires, so neither could field large numbers of airplanes. Germany had the industrial capacity, but concentrated its airplanes on the decisive Western Front. The long distances placed a priority on long-range reconnaissance missions, and the low density of combat airplanes and the lower priority of the Front for Germany resulted in a much less intense air combat environment than the West.

Older airplane types lasted longer in service on the Eastern Front, and second-rate French and German airplanes, like the SPAD SA series, Roland D.II, and Aviatik C.II, were often assigned there, where they could be of use; first-rate airplanes were frequently reserved for the more demanding Western Front. On the other hand, weather extremes meant airplanes had to be robust and reliable.

Surprisingly, Russia had the world's largest air force at the start of the war, but the small Russian aviation industry had limited production capacity and could not keep pace with attrition. This made Russia largely dependent on France for airplane designs built under license, for completed airplanes, and especially for aero engines. Inadequate supply of airplane engines was a chronic problem for Russia, which was able to produce only 1,511 aero engines during the entire war. Engines from downed airplanes, both Russian and enemy, were eagerly salvaged for re-use in Russian airplanes.

In his book *Memories and Impressions of War and Revolution in Russia 1914–1917*, Russian General Basil Gruko, former commander of the Russian 5th Army, sums up the Russian aviation experience: "Of course the Germans had far fewer machines on our front, taking its extent in consideration, than they had on their Western Front, which showed their

Above: At 44 years old , *Lt.* Friedrich N. Güttler was one of the oldest fighter pilots and was, of course, known as 'Papa' to the other pilots of his unit, *Jasta Ober Ost* on the Eastern Front. Here he is in the cockpit of his Roland-built D.IIa. Second-rate aircraft like the Roland D.II were often relegated to the Eastern Front where they could still be useful in the less intense air combat environment there.

Facing Page, Bottom: Capt. Paul d'Argueff was the third-ranked Russian ace with 15 victories.

Left: The majority of WWI combat aircraft produced in Russia were license-built French designs; here a Dux-built Nieuport 16 fighter. Moreover, most aero-engines built in Russia in WWI were license-built French designs. At the time Russia had a small industrial base and could only build 1,511 aero engines during the entire war.

Left: Igor Sikorsky was one of the leading Russian aircraft designers during WWI and is best known for his Ilya Mouromets bomber designs, which were the most advanced in the world at the time. However, he produced many other designs, including the S-16, the first aircraft designed in Russia that could be used as a fighter. The S-16 had a wide cockpit that could seat two for training or light scouting, and either skis or wheels could be fitted as needed. Armed with a machine gun, it became a first generation fighter.

Left: Developed from the the Sikorsky S-16, the S-20 was the most advanced Russian-designed fighter to be produced, but only five were built before production was terminated to enable the factory to build more Ilya Muromets reconnaissance bombers, which were more critical to the war effort. Power was either a 100 hp Monosoupape or 120 hp Le Rhône.

Above & Right: Although the great majority of WWI combat aircraft used in Russia were of French origin; Great Britain also made a small contribution. Here are two photos of a Vickers FB.19 Mark I in Russian service, one of six supplied to the IRAS.

need of having aviation power equal to our Allies' flying forces. Nevertheless, the Germans were always able to have superiority over us in Aviation". Similarly, Russian General A.A. Brusilov, in his book *A Soldier's Note Book*, states "during the whole of the war, we were badly handicapped by the shortage of machines... As for dirigibles, we had none, for one cannot take seriously the few specimens that we had bought, at a high price, from foreign countries; they were out of date, inefficient, and of no practical value. Altogether it must be recognized that, compared with that of the enemy, our technical equipment was quite inferior." Although the Austro-Hungarian air service had at best marginal superiority over the Russians, German technical superiority made a difference in the East.

Above & Right: Roland D.IIa(Pfal) 374/17 at Koledziany airfield on the Eastern Front. In addition to the standard insignia, the lack of a tail skid fairing identifies this aircraft as Pfalz-built. The Roland D.IIa was used extensively on the Eastern Front, where the air combat environment was more permissive than on the Western Front and the D.IIa could make use of its strengths, good speed, strong armament, and a robust structure, against two-seaters while its weaknesses, poor maneuverability and modest climb over 3,000 m, were not as important.

Above: Spad SA.4 on skis serving with the IRAS.

Left: The 19th Corps Detachment lined up under threatening skies. The skull and cross bones on the rudders of these fighters were the unit insignia. Morane-Saulnier I and Nieuport fighters are visible.

Below: A later view of the 19th Corps Detachment with later-model Nieuport fighters; the Moranes are gone and Spad 7s are visible. The unit insignia could be black on white or white on black.

Above: Two of the six Roland D.III fighters supplied to Bulgaria on Levunovo airfield in 1918; the late-style insignia verify the date. The Roland D.III was quickly replaced in production by the superior Pfalz D.III.

Above: Bulgarian Roland D.III that has been crashed, probably the result of another landing accident. The Roland D.III was only used on the Eastern Front; it did not have the performance or maneuverability to compete on the more demanding Western Front. Pfalz had built Roland D.I and D.II fighters under license and developed their own original Pfalz D.III based on what they had learned from the Roland designs.

Above: Bulgarian Roland D.II in flight.

Above: Bulgarian airmen in front of a Roland D.III of the Bulgarian air service in 1918. This aircraft has a 160 hp Mercedes D.III engine. Some Roland D.III fighters had the 180 hp Argus As.III.

The Middle East

From an aviation perspective the Middle East was an interesting backwater during the war. The Ottoman Empire had very limited transportation and industrial capacity and was unable to manufacture either airplanes or aero-engines, so was totally dependent on Germany. Germany, preoccupied with the decisive Western Front and faced with long, insecure supply lines to the Ottoman Empire, was unable to provide much help. A major motivation for Britain to attack the Ottoman Empire was its strategic location, which blocked an easy southern sea route to supply Russia and also threatened the important Suez Canal. When Bulgaria entered the war allied with Germany and the Ottoman Empire, a direct railroad line between Germany and Turkey became available; transportation was now much easier but Germany could spare few aircraft.

In the late stages of the Gallipoli campaign four German fighter pilots arrived with their Fokker *Eindeckers*; Buddecke shot down four airplanes, Meinecke two, Schüz two, and Muhra one, to give the Allies a hard departure.

New German units *Fliegerabteilungen* 301, 302, 303, and 304b, arrived in Aleppo in October 1917 and were immediately sent to Palestine. Each *Abteilung* had two Albatros D.III fighters and a small number of AEG C.IV reconnaissance planes.

Meanwhile, more modern airplanes, including Halberstadt fighters, Albatros D.II and D.III fighters (equipped with two radiators for the heat) had been delivered to the Ottoman air service. In early 1918 20 Albatros D.V fighters were delivered, but these had more than usual structural defects and after two fatal crashes due to wing failure, one the *Staffelführer*, they were grounded.

German aviation reinforcements were countered by delivery of RE.8 and SPAD 7 aircraft to the British in even greater numbers, plus such rare types as the DH.1a, Bristol M.1c, and Vickers FB.19. Finally DH.9's and SE.5a and Bristol F.2b fighters arrived. Flown by 40th (Army) Wing (No. 1 Squadron, Australian Flying Corps, and No.111, No.144, and No.145 Squadrons, RAF) and 5th (Corps) Wing (No.14, No.113, and No.142 Squadrons, RAF) in Palestine, these aircraft outnumbered the Turkish and German airplanes and gave Britain both quantitative and qualitative air superiority.

Despite Allied superiority, the low density of airplanes over the front enabled German and Turkish units to engage in bombing, artillery spotting, and long-range reconnaissance around Palestine. Eight German reconnaissance airplanes arrived in May 1918; these were a combination of LVG C.V and Rumpler C.IV two-seaters that enabled continuation of reconnaissance missions even in the face of modern British fighters. The next month a final delivery of eight German fighters arrived; sources differ as to whether these were Albatros D.III or Pfalz D.IIIa machines. While there was frequent air combat with losses on both sides, much of the attrition was due to accidents and mechanical wear and tear. By July the British, aided by a great numerical advantage, held complete air superiority over the Palestine battlefield and in August German flying units were running out of airplanes.

On 19 September 1918 the British launched the decisive Battle of Megiddo and the Turkish front was broken. The Jenin airfield was singled out for special attention; No.111 Squadron RAF in their SE.5a's kept continuous air patrols over it and strafed it to prevent reconnaissance missions from taking off.

British forces advanced through the gap and, supported by RAF air attacks on retreating Turkish troops, routed the Turkish army. On 20 September British cavalry, still useful in this theater, suddenly overran the Pascha units' airfield at Jenin, capturing several German aircraft.

The AFC and RAF had been watching five

Above: Buddecke stands near Fokker E.III 96/15 in Turkey accompanied by two other servicemen, *Oblt.* Pommerer (left) and *Kptlt.* Meyer (center). (courtesy of Tobias Weber)

Above: *Lt.* Hans Henkel flew this Pfalz E.II, the only one in the Middle East, with *Flieger Abteilung* 300 "Pascha."

Above: *Lt.* Emil Meinecke poses in front of his Halberstadt D.V fighter after a hard combat with British aircraft. Meinecke flew with the fighter element of Ottoman *Flieger Abteilung* 6 at Chanak (now Cannakale) in the Dardanelles and scored six victories to become an ace. His Halberstadt wore Turkish insignia.

Right: The dumpy Vickers FB.19 Mark II fighter lacked the performance to be used on the more demanding Western Front but was used in small numbers in the Middle East. Its single gun is visible on the fuselage side.

Top: *Lt*. Emil Meinecke's Albatros D.II in Turkish markings; his mechanic stands in front of the aircraft.

Above: Albatros D.III fighters of *Flieger Abteilung* 300 "Pascha" mounted two radiators to handle the desert heat.

Left: *Lt*. Emil Meinecke in an Albatros D.III with twin radiators fitted for the hot Middle East climate.

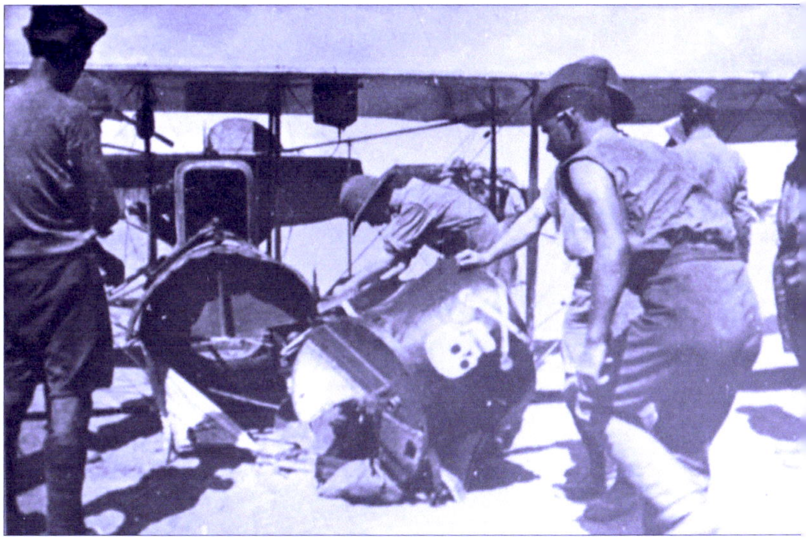

This Page: The modest performance of the DH.1A limited its use to the "sideshows" and training squadrons. No.14 Squadron was probably the only unit to use the DH.1A operationally during its service in the Middle East. 4609 and 4610 are known to have served with this unit. The skull and cross bones marked DH.1A may have been 4609, which was finished in a similar scheme, but the skull insignia can not be confirmed as being applied to this aircraft. The crew were reportedly killed in the crash circa 1917. The Lewis gun had a simple spigot mount. The high radiator distinguished the DH.1A's water-cooled 120 hp Beardmore engine.

strategic hill passes the Ottoman army would have to retreat through. On 21 September the RAF and AFC caught the Turks retreating though Wadi Fari'a and devastated them with nearly six hours of continuous bombing and strafing attacks, causing immense casualties and complete panic. The airmen called the wadi "The Valley of Death" and these attacks broke the morale of the Ottoman troops. The airmen were both proud of their accomplishment and sickened by the slaughter. On 1 November the Turks signed an armistice.

Left: No.1 Squadron, AFC; the officer in front holding the stick is Richard Williams, later Air Vice Marshall Sir Richard Williams of the RAAF.

Below Left: DH.2 and Vickers F.B.19 of No.14 Squadron, RFC, in the Middle East.

Below: No. 72 Sqdn., RAF, in the stark terrain of the Middle East. The flight of Bristol M.1c monoplanes is distinctive; due to a prejudice against monoplanes the M.1c was used operationally only in the Middle East. It was also used for training in the UK.

Above: On 17 February 1917 FSL G.T. Bysshe in Bristol Scout 8996 was shot down by *Off.Stv.* Emil Meinecke and captured. The Scout was repaired, including installation of an Oberursal rotary from a Fokker E.III, marked in Ottoman colors, and flown by Meinecke! The photo above shows it after modification; see the back cover for a profile.

Right & Above Right: The Bristol Scout after engine failure and a crash landing while being flown by Meinecke.

Right: One of the very few actual color photos from WWI is this Bristol F2B of No.1Sqdn. AFC. A large portion of the fabric is white to reflect the sun. Lt. Lew Potts and Lt. James Traill of No.1 Sqdn. AFC discuss their reconnaissance in front of F2B A7194. No. 1 Sqdn. was a reconnaissance squadron but produced some aces. This Frank Hurley color photo was taken at EL Mejdel aerodrome in February 1918. The F2B was the best warplane in the Middle East and enabled the British to achieve air superiority. (AWM)

Top: Bristol M.1c A5142 of No.67 Sqdn. at Deir-el-Belah.

Above: Fast Bristol M.1c A5141 was used by No.67 Sqdn. at Deir-el-Belah. No.67 Sqdn. became No.1 Sqdn., Australian Flying Corps (AFC) on February 6, 1918.

Above: This Bristol M.1c was used by C Flight, No.72 Squadron in Mesopotamia in 1918.

Right: Bristol FB.19 Mark II A5236 of No.67 Sqdn., RFC, at Deir-el-Belah. Unfortunately, the men are not identified.

Pfalz D.IIIa in Turkish Service

A batch of late production Pfalz D.IIIa fighters, probably 20 aircraft, was built for Turkey. At right is a photograph of one of these aircraft with Turkish insignia, the crescent and star insignia, not the black square used by German aircraft in Turkey.

Below, Pfalz D.IIIa 1365/18 is shown being shipped to Palestine for Turkish use. The rudder has been painted white for the *Balkenkreuz* insignia, and an unbordered *Balkenkreuz* is on the fuselage. These photos show the last Pfalz D.IIIa series went beyond D.1349/18; probably 20 Pfalz D.IIIa 1350–1369/18 were built for Turkey.

Above: Pfalz D.IIIa D.1350/18 enroute to Turkey in German markings and typical Pfalz silver-gray finish.

Top Right: Pfalz D.IIIa with Turkish crescent and star insignia.

Above: Pfalz D.IIIa fighters at Kara-Pumar on their way to Palestine. An upper wing is mounted on the flat car behind the tail of the D.IIIa on the extreme right. The wing is covered in lozenge fabric with a wide-bordered *Balkenkreuz*.

Above: Martinsyde Elephants and F.2B fighters in Palestine are shown in this extremely rare original color photograph. (AWM)

Above: Martinsyde G.102 7488 flown by Capt. A.W. Ellis, No.67 Sqdn., RFC.

Right: Bristol Fighter and RE.8s of No.1 Sqdn., AFC; this unit had been No.67 Sqdn., RFC until Feb. 6, 1918.

Above: Martinsyde Elephants and F.2B fighters in Palestine are shown in another extremely rare original color photograph. (AWM)

Fighters Weird & Wonderful

The vast majority of fighters flew over the battlefronts by day and engaged in air combat with other aircraft. But there were circumstances requiring fighters with special capabilities, such as combat at night, intercepting Zeppelins, armored fighters for fighting over the trenches, or the ability to operate from water or from ships, and these unusual fighters will be discussed here.

Night Fighters

Above: This Sopwith Camel has been converted to 'Comic' configuration for night fighting. The guns have been moved above the upper wing so their muzzle blast will be blocked from the pilot's view by the wing, preserving his night vision, and the cockpit has also been moved aft.

The growing practice of night bombing generated a demand for night fighters. Night fighters were especially important to Britain because Britain was the target for German strategic night-bombing campaigns throughout the war, first by Zeppelins and later joined by twin-engine Gotha bombers and multi-engine Giant bombers.

During the early war years a wide variety of primitive aircraft were used to try and intercept Zeppelins. These aircraft generally had marginal performance and posed only a strictly limited threat to airships. To remedy this problem a number of eccentric aircraft were designed specifically as

Above: Before the invention of radar, British soldiers man a detector to locate enemy aircraft at night by their sound.

Above: Camel B2402 of No.44 (HD) Squadron modified to 'Comic' configuration for night fighting.

Right: Standard configuration Camel used for night fighting by No.44 (HD) Squadron. The aircraft has been painted black overall and the national insignia colors have been toned down for reduced visibility at night.

Zeppelin hunters during the first half of the war; these aircraft were designed to intercept the high-flying Zeppelins at night and attack them with various flexible, upward-firing weapons firing incendiary ammunition. Great endurance for long standing patrols was a key requirement. The Zeppelin hunter designs were characterized by being multi-seat aircraft, with one or two gunners firing flexible machine guns with an extensive field of fire upwards due to the airships higher ceiling. The recoilless Davis gun was planned as armament for some of these aircraft in place of machine guns, but the Davis gun was not used operationally against airships. The biggest limitation of the Zeppelin hunter designs – and other early aircraft used for night fighting – was their inadequate speed, climb, and ceiling to enable them to close with an airship once sighted.

As aircraft performance improved, the Zeppelins were also improved, especially their ceiling and climbing ability. The net result of these demands on night fighter performance was to eliminate eccentric designs with multiple gunners in favor of the best fighters developed for day fighting with their superior speed, climb, and ceiling. Most of these aircraft were modified slightly for night operations by such additions as lighted flight instruments, flares to light the landing area, etc. The Sopwith 1½ Strutter and its successor in night fighting, the Sopwith Camel, were used in both standard models and extensively modified variations to optimize them for night fighting. However, even in the case of these two types, the standard aircraft with additional night-flying equipment were more numerous.

Above: FE.2d 1882 and 1882 assigned to No.78 (HD) Squadron for night fighting as part of the defense of London and the southern approaches.

Left & Below: Sopwith 1½ Strutter two-seaters assigned to No.78 (HD) Squadron for night fighting as part of the defense of London.

Right: This BE.2d assigned to No.36 (HD) Squadron for night fighting carries Le Prieur rockets for anti-Zeppelin patrols. The most common British night fighter during the middle war years was the BE.2c. The observer's cockpit has been faired over.

Above: This experimental FE.2b conversion of A781 was fitted with a searchlight for use as a night fighter as shown. Two Lewis guns were mounted along side the searchlight which, along with four landing lights under each wing, were powered by a wind-driven generator under the nose.

Below: Bristol FB.2b "1" and "2" assigned to No.36 (HD) Squadron for night fighting. The Bristol Fighter was one of the most effective aircraft converted to night fighter duties.

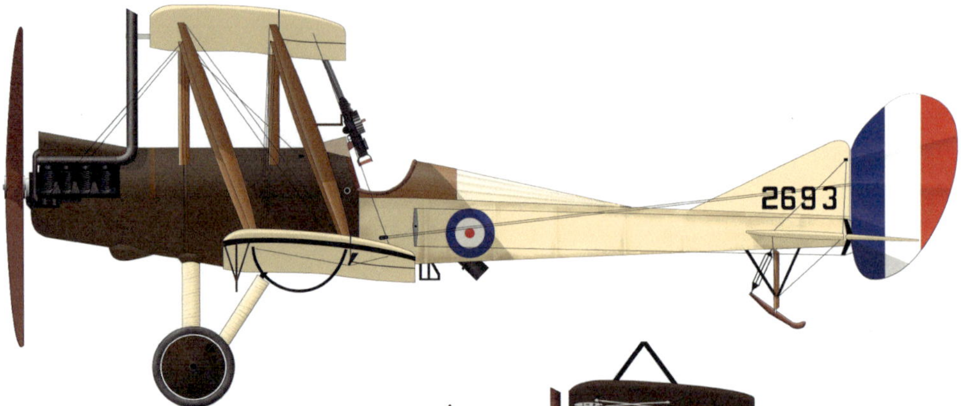

BE.2c 2693 of Lt. William Leefe Robinson, No.39(HD) Sqdn., 2 Sept. 1916 when Robinson destroyed SL.11, for which he was awarded the Victoria Cross.

BE.2e of No.50(HD) Sqdn., Detling Aerodrome. The unit insignia was painted under the upper wing.

BE.12 6655 of No.77(HD) Sqdn., June 1918

BE.12a A582 of No.50(HD) Sqdn., May 1917

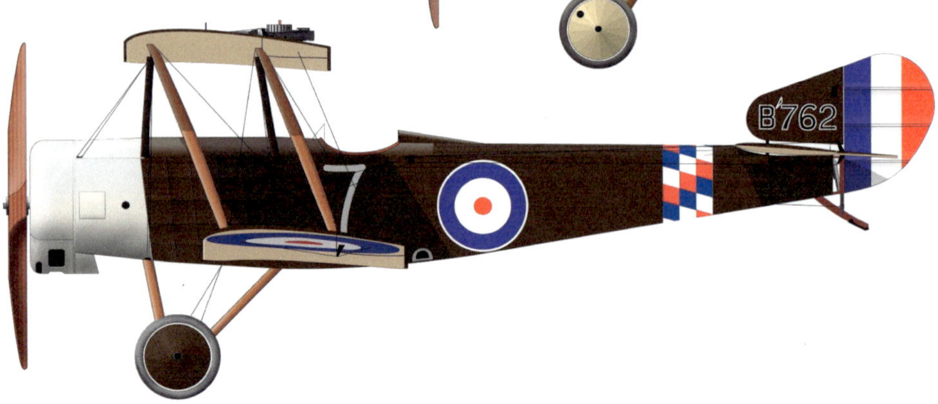

Sopwith 1½ Strutter 'Comic' of No.78(HD) Sqdn., Martlesham Heath Aerodrome, November 1917

Right: BE.12b C3094 was assigned to No.76 (HD) Squadron for night fighting. This aircraft was flown by Lt. R. MacDonald. Armament was an over-wing Lewis gun that is not fitted in this photo; just the mount is visible. The BE.12 series was a single-seat fighter developed from the BE.2 series two-seat reconnaissance aircraft, but was almost useless during daylight due to its excessive stability and moderate performance. A number were employed as night fighters, where their stability and long patrol endurance were assets.

Left: Even the Avro 504 trainer was modified for use as a night fighter; here Avro 504K E3273 and E3278 assigned to No.77 (HD) Squadron for night fighting are preparing to take off.

Below: BE.12bs and BE.12s assigned to No.77 (HD) Squadron for night fighting. lined up at Penstone.

Above: Sub-Lieutenant Reginald Alexander Warneford was the first pilot to destroy a German airship in flight and was only the second British aviator to be awarded the Victoria Cross. His feat proved the Zeppelins were vulnerable to interception and destruction – if they could be found – and greatly reduced the anxiety about the Zeppelins. He pursued *LZ.37* for more than an hour before destroying it by air-to-air bombing.

Above & Below: On 6 June 1915 Morane-Saulnier L 3253 was used by Flt. Sub-Lt. Reginald Alexander Warneford to destroy Zeppelin *LZ.37*, the first Zeppelin to be destroyed in flight by an aircraft. Serial 3253 was armed only with bombs and Warneford downed *LZ.37* by air-to-air bombing. On 17 May 1915 *LZ.39* had easily out-climbed Warneford who had attacked it while flying a Voisin. A pre-war design, the Morane-Saulnier L was powered by an 80 hp Le Rhône; it had a maximum speed of 77 mph and a ceiling of more than 11,000 feet.

Zeppelins were very much on British minds before and during the war. Britain had been secure from enemy attack since the defeat of the Spanish Armada in 1588, and the Zeppelin had the range, endurance, and bomb load to change that. Desperate to win the war after the failure of the Schleiffen Plan resulted in stalemate on the battle field, Germany resorted to bombing Britain with the hope of driving Britain out of the war. With the limited range and bomb load of the early-war airplanes, that left airships as the weapon of necessity.

Filled with highly flammable Hydrogen for lift, airships were extremely vulnerable. They could withstand numerous bullet holes, but once a fire started, an airship was doomed. However, to destroy an airship, another aircraft first had to come within effective weapons range, and that was the first key problem that had to be solved.

Effective weaponry was the second key problem that had to be solved. Recognizing this, a great deal of effort was invested in long-range weapons firing incendiary ammunition to set the airships on fire.

The first airship to be destroyed by an aircraft in flight was *LZ.37*; on the night of 6 June 1915 Sub-Lieutenant Reginald Alexander Warneford was able to stalk the airship and dive down from above, dropping bombs as he did so. The airship was consumed in the resulting fire and Warneford, who flew a Morane-Saulnier L on that mission, was awarded the Victoria Cross. However, Warneford was only able to position his Morane above *LZ.37* because it was returning to its base in Belgium and was losing altitude in preparation for landing.

After Warneford's victory over *LZ.37*, the next notable British success came more than a year later. On 2 September 1916 Capt. William Leefe Robinson, flying a BE.2c armed with a single flexible machine gun, downed *SL.11* near London by using incendiary ammunition. Again, the pilot was awarded the Victoria Cross for his accomplishment.

Robinson's victory was significant militarily and culturally. First, his victory was the first time an airship had been brought down with the standard weapons then in widespread use, a typical night fighter and machine gun with incendiary ammunition. Furthermore, the airship was not caught at a disadvantage descending to land but was engaged in its typical mission profile when destroyed. Finally, the long-suffering civilian population of London was able to see the airship burn all the way to the ground as the fierce fire was visible for 60 miles, a range that included most of London. This brilliant success near the capital seemed to eliminate most of the civilian fear of the airship as an effective weapon, reducing it to the status of dangerous nuisance for most of the public.

Robinson was later downed and made POW leading the first Bristol Fighter combat sortie.

Above: Capt. William Leefe Robinson was the first pilot to destroy a German airship over Britain and was the eighth British aviator to be awarded the Victoria Cross. *SL.11* fell burning fiercely within view of London, and this sight of a burning airship apparently broke the fear of Zeppelins for good. Robinson's feat greatly encouraged a public shocked by the losses on the Somme and tired of being bombed by German airships.

Below: The BE.2c was one of the most widely-used night fighters during the mid-war years, and on 2 September 1916 William Leefe Robinson, shown here in the cockpit of BE.2c #2693, shot down Shütte-Lanz airship *SL.11* flying this BE. This was the first airship brought down over Britain by a night fighter and Robinson was immediately awarded the Victoria Cross for this accomplishment. Here the ground crew are displaying the center section that was damaged by Robinson's own gunfire while attacking *SL.11*.

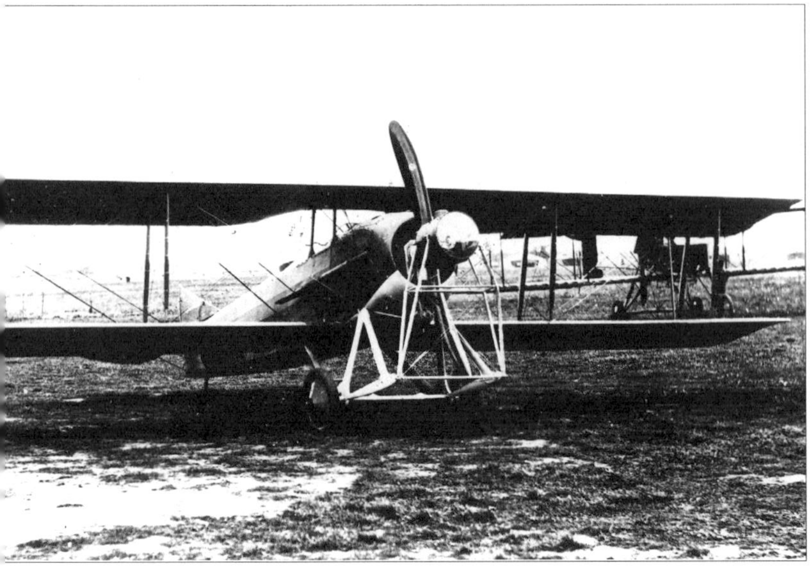

Above & Below: The Spad 11 was designed as a two-seat fighter, but its mediocre maneuverability resulted in it being utilized as a reconnaissance aircraft. Nevertheless, one Spad 11 was fitted with a searchlight for use as a night fighter as shown here. The installation remained experimental. Use of searchlights to illuminate target aircraft at night could give more advantage to the target aircraft, whose gunners could clearly see the aircraft shining the searchlight.

Above: Derived from the Breguet 14 A2 reconnaissance plane that it very closely resembled, the Breguet 17 C2 was a two-seat, long-range escort fighter and night fighter. Powered by a 400 hp Renault, it had two fixed guns for the pilot, a two-gun Lewis turret for the gunner, and a third Lewis mounted to fire downward for the gunner's use. First flown in mid-1918, only ten were built because the Hanriot HD.3 was considered a superior fighter.

Above: The SE.5a was effective as a night fighter as well as a day fighter. Here Capt. Insall of No.50(HD) Squadron is ready for take off in his SE.5a. Primitive flame dampers have been added over the exhausts to preserve the pilot's night vision and render the aircraft less visible to its targets, and the markings also have been toned down.

Above: Nearly all Sopwith Camel night fighters were based in the UK for home defense. However, in 1918 No.151 Squadron was sent to France to intercept German night bombers, a deployment made practical because Germany shifted its night bombing efforts to the Western Front to help the army. No.151 Squadron downed two Giant bombers during their time in France.

Above & Below: The unusual Belgian single-seat GN3 night-fighter mounted two Lewis guns and was powered by a Hispano-Suiza V-8 engine. The radiators were mounted on the sides of the nacelle and a searchlight with 30 cm reflector was fitted to the front of the nacelle to illuminate the target. Of course, using the searchlight would also give a defending bomber gunner a good target to shoot back at! Together with its generator/dynamo, the searchlight weighed 100 kg and cost 4,000 francs; it was effective up to a range of 1,500–2,000 m. The GN3 was likely flown on operations by Prosper Georges during attempts to intercept Gotha night bombers. Only one GN3 was built.

Zeppelin Hunters

Early in the war, Germany's Zeppelins were an obsession for both sides due to their ability to directly attack Great Britain, a threat that had last been experienced when the Spanish Armada attacked in 1518. At the beginning of the war the Zeppelins had the ability to out-climb aircraft by dumping ballast and could fly higher and almost as fast. Coupled with their use at night, the Zeppelin threat posed unique defensive challenges for the British in those times decades before the invention of radar.

Zeppelin interceptors had to be night fighters of long endurance to fly area patrols and pursue the attackers over long distances. Upward-firing armament was important because the Zeppelins would often fly higher than the aircraft and could, as noted, out-climb aircraft by dumping ballast. The result was a number of eccentric designs, often with gunners manning flexible guns with good upward fields of fire.

Zeppelin interceptors were a British exclusive; with the German Army on her soil, France had little energy to spare for countering marginally-effective weapons like Zeppelins. On the other hand, concern about popular opinion made the Zeppelin an important weapon in psychological terms that the British had to deal with.

The Zeppelin interceptors were all aircraft of mediocre performance and none were placed in production. The real solution to the Zeppelin threat lay in conventional aircraft of higher performance coupled with better defensive strategy and tactics.

Above: The Sopwith L.T.R. Tr. was created as a multi-seat, long-range escort and anti-airship fighter. Designed to the same requirement as the Vickers F.B.11 and Amstrong Whitworth F.K.12, like them it was powered by a single 250 hp Rolls-Royce V-12. Also like the F.B.11 and F.K.12, the L.T.R. Tr. had two gunners with flexible Lewis guns and remained a single prototype. One gunner sat behind the pilot and the other occupied the streamlined nacelle built into the top wing. Designed before the availability of practical interrupter gear, it was too slow for its intended role.

Above: The N.E.1, for Night-flying Experimental, was a 1917 night fighter design powered by a 200 hp Hispano-Suiza. It mounted a searchlight in the nose for target illumination, and armament was either a Coventry Ordnance Works quick-firing gun or a Vickers rocket gun, supplemented by Lewis guns.

Right: The Vickers F.B.11 was designed as an escort fighter and anti-Zeppelin interceptor. Powered by a 250 hp Rolls-Royce V-12, the nacelles over the wing housed a gunner with flexible gun and another was seated behind the pilot.

Below: The Armstrong Whitworth F.K.12 anti-Zeppelin interceptor in its second form. Powered by a 250 hp Rolls-Royce V-12, the nacelles on the wings each housed a gunner with flexible machine gun.

Below: As might be guessed from its configuration, the Robey-Peters Gun-Carrier was designed for the same roles as the Vickers F.B.11 and Amrstrong-Whitworth F.K.12, that is, escort fighter and anti-Zeppelin interceptor. Like those types, it was also powered by a 250 hp Rolls-Royce V-12. The nacelles under the wing each housed a gunner with flexible Davis recoilless gun (the 2-pounder Davis guns were not fitted) and the pilot was seated well aft of the wing. The first aircraft (serial 9498 shown here) crashed on its maiden flight in May 1917 and the second aircraft ordered was not completed.

Above: The Supermarine Night Hawk was designed by Flt.Lt. Pemberton-Billing as an anti-Zeppelin interceptor and shared many attributes with the earlier P.B.29 that was designed for the same role. Both aircraft were twin-engine quadraplanes designed for long endurance night patrols. The Night Hawk had a flexible 1½-pounder Davis gun mounted above the cabin firing forward to attack the airship and two flexible Lewis guns, one aft the Davis gun and one in the nose, for defense. A searchlight was mounted in the nose to illuminate the target at night and secondarily illuminate the landing field at night. Powered by two 100 hp Anzani engines, the aircraft had the remarkable endurance of more than 18 hours.

The conventional night fighters, Bristol Fighters, DH.4s, SE.5as, and Camels were the aircraft that successfully defeated the Zeppelin threat and forced the Germans to switch to twin-engine Gotha and multi-engine Zeppelin-Staaken Giant bombers to continue their strategic bombing offensive against Great Britain.

Below: The A.D. Scout, or Sparrow, was designed in 1915 as an anti-Zeppelin fighter. Powered by an 80 hp Gnôme rotary and intended to be armed with a Davis recoilless quick-firing gun, it was tricky to fly and quickly abandoned.

BE.12b

FE.2b A5635 of No.33(HD) Sqdn., Kirton Lindsay Aerodrome, November 1917

FE.2b A5548 of No.51(HD) Sqdn., Early 1918

FE.2b A5729 single-seat conversion of No.51(HD) Sqdn., Oct. 1917–April 1918

FE.2b D9124 single-seat conversion of No.51(HD) Sqdn., Summer 1918

Convoy Escort Fighters

The multi-seat 'convoy' escort fighter concept started early in the war and lasted into at least 1943, when the USAF tried it in combat with the B-40 (a variant of the B-17) and the B-41 (a variant of the B-24). Of all the convoy escort fighters conceived or built, only one was successful in the role for which it was devised; the Caudron R.11.

The Caudron R.11 was a fast, heavily-armed twin-engine aircraft designed for reconnaissance and as a long-range, multi-seat escort for French bombers, in the event the Breguet 14 B2. Powered by two Hispano-Suiza engines of either 200 hp or 215 hp, the three-seat R.11 had gunners fore and aft each armed with twin Lewis guns on a rotating turret. In addition, a fifth, downward and rearward-firing Lewis operated by the forward gunner protected the R.11 from attacks from below and behind.

Heavily-armed and as fast as single-seat fighters, the Caudron R.11 escorted daylight bombing raids by Breguet 14 B2s and blasted a path for the bombers during 1918. The limited number of German fighters were unable to defeat the large French formations of bombers, themselves well-armed, protected by Caudron R.11 escorts.

Above & Below: The Caudron R.11 was the only aircraft ever successful as a multi-seat escort fighter.

Above: If the Caudron R.11 was way to be successful as a multi-seat escort fighter; the eccentric Breguet 11 *Corsaire* was not. This beast had three 220 hp Renault engines, four crewmen, and was intended to be armed with a 37mm Hotchkiss cannon and three machine guns, with bombs an alternative armament. It was not chosen for production.

Caudron R.11 escort fighter of R.46, Summer 1918.

Bristol F.2b C4645 "4 Diamond" of No.39(HD) Sqdn., 1918

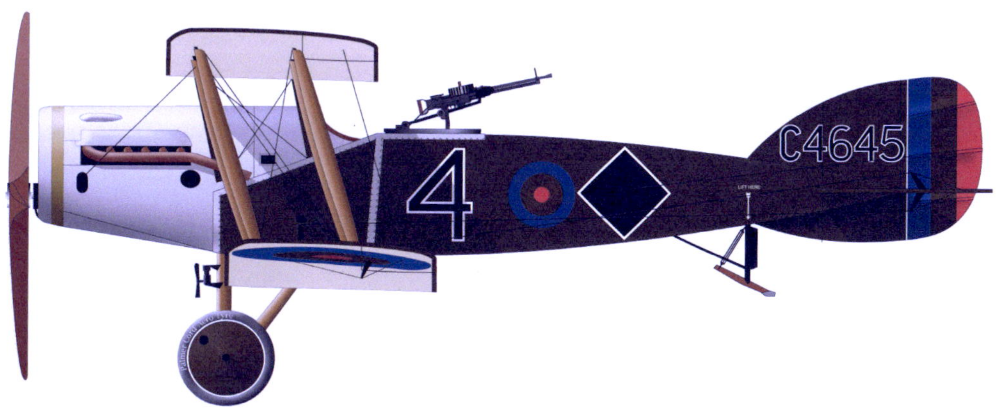

Sopwith "Comic" Camel of No.51(HD) Sqdn., 1918

Sopwith "Comic" Camel E5165, 1918

Armored Fighters

The relatively low-powered engines available during most of the war made armored warplanes of the time too heavy and sluggish for most practical combat roles. A few aircrews added armor plates to their individual aircraft, and small numbers of armored Caudron G.4 and Farman F.43 aircraft were used, but the performance of these aircraft suffered from the weight of their armor and few were used in combat. The first truly armored production aircraft to see combat in numbers were the German J-type infantry aircraft that started appearing over the Western Front in the summer of 1917. These cumbersome aircraft assigned to infantry cooperation relied on their armor protection rather than their very modest performance to survive at low altitudes over the front.

As the air war developed, ground-attack against enemy troops, machine guns, artillery, and tanks became more important, and anti-aircraft weapons and tactics became more widespread and effective in response. These developments, coupled with the greater power of available engines, forced the air services to again consider armored aircraft for low-level operations. Both the British and German air services looked for an armored fighter for the related roles of ground attack and interception of low-flying aircraft such as the J-types, and both developed armored fighters for these roles.

Britain Builds the Sopwith Salamander

The first British attempt at an armored fighter was the Sopwith T.F.1 [Trench Fighter 1], a Camel experimentally fitted with downward-firing Lewis guns and some rudimentary armor around the cockpit to serve as a ground strafing machine; it did not see production. The Sopwith T.F.2 Salamander, named for a creature that – in mythology – lived in fire, was developed from the Sopwith Snipe fighter that reached the Western Front in October 1918. The Salamander used the same 230 hp Bentley B.R.2

Above Left: Photograph of the armor for the Sopwith TF2 (Trench Fighter 2), later designated the Salamander.

Above: The Sopwith Salamander used the flying surfaces and engine of the Sopwith Snipe; its flat, armored fuselage sides differentiated the Salamander from the Snipe.

Left: Special camouflage was developed for the Salamander due to its intended low-altitude operating environment.

Above: Derived from the earlier Vickers F.B.26 fighter of 1917 powered by a 200 hp Hispano-Suiza, the F.B.26A of 1918 was an armored fighter prototype powered by the 230 hp B.R.2 rotary engine. With 500 pounds of armor fitted inside the nacelle, both pilot and engine were protected. Armament was two fixed Lewis guns. Despite excellent performance for an armored pusher, the F.B.26A was abandoned in favor of the Sopwith Salamander, perhaps because the Salamander shared so many components with the Snipe fighter then in mass production.

rotary engine and carried the same armament of two Vickers machine guns used by the Snipe. The two-bay wings and the tail surfaces were very similar, and the entire forward fuselage of the Salamander was constructed of 650 pounds of armor plate. The main feature that distinguished the Salamander was its flat-sided fuselage, whereas the Snipe had a rounded fuselage.

Because it was a direct development of the Snipe, the Salamander suffered from the Snipe's prolonged development of its flying surfaces. As a general-purpose fighter the Snipe also appears to have had higher priority. Regardless, the Salamander had good performance at low level and was ordered into production in the summer of 1918, months before the Snipe reached the front. The Salamander's production was not rushed, and by the end of October, just days before the Armistice, the RAF had 37 Salamanders, of which only two were in France. On that same date there were 97 Snipes in France and the type had been in combat for several weeks.

German Armored Fighter Prototypes

No armored fighter was placed into production by Germany, but several experimental designs were built and tested.

First was the AEG DJ.I Triplane. The engine, fuel, and pilot were protected by a steel shell forming an integral part of the airframe and the rear fuselage was of aluminum. This was designed during the height of Idflieg's triplane craze and the aircraft had triplane wings with AEG's customary steel-tube spars and wooden ribs. The test pilot reported the triplane as easy to fly but not particularly fast. Work on the AEG DJ.I triplane was stopped in May 1918 because of "poor flight characteristics," perhaps related to the thin airfoil section used for the wings.

Disappointed by the mediocre flight characteristics and low speed of the AEG DJ.I triplane, AEG engineers brought out a biplane version, the AEG DJ.I fighter, powered by the same experimental 195 hp Benz Bz.IIIb V-8 engine used in the earlier triplane. The biplane use the same armored fuselage as the triplane coupled to thick biplane wings unlike those on any other AEG design of the war. The wireless wing cellule was designed to minimize damage from ground fire.

Converting to a biplane wing cellule improved the aircraft's speed and climb rate and likely its handling qualities due to its thicker airfoil. A second AEG DJ.I powered by a 245 hp Maybach Mb.IVa engine was under construction from May to July 1918 and probably performed some flight trials. A third prototype was reported under construction in September but it is not known if it was completed.

Perhaps not surprising given the great number of fighter prototypes built by Fokker, the third German armored fighter was the Fokker V37 monoplane. Derived from the earlier Fokker V27, and like it powered by a 195 hp Benz Bz.IIIb liquid-cooled V-8 engine, the Fokker V37 parasol monoplane was an armored fighter designed for low-level fighting and

Above: The AEG DJ.I triplane was envisioned as a low-altitude fighter armored to withstand ground-fire as it intercepted enemy ground-attack aircraft at low altitude, a role similar to that foreseen for the Sopwith Salamander. It was powered by a 195 hp Benz Bz.IIIb V-8 engine, which did not reach production in time to power any warplane over the front.

ground attack and bore a strong family resemblance to the production E.V/D.VIII monoplane fighter.

Clearly the radiator had to be protected from ground fire to make an armored airplane with water-cooled engine survivable, and the V37's solution was clever. The radiator was mounted behind the propeller and protected by a disk of armor. The propeller was given a large spinner for streamlining, and air was forced through the radiator by six fan blades mounted in the plane of the propeller. The entire forward fuselage, including the engine and cockpit, was protected by steel armor plate 2.5mm thick. The armor behind the pilot even extended up behind his head and was faired into a headrest.

Had the V37 design progressed, it would certainly have been fitted with the standard fighter armament of two synchronized machine guns. However, the V37 arrived too late to be properly evaluated, let alone produced in quantity, so its performance and flying qualities are not known.

Armored Fighters in Retrospect

The low power available to WWI aircraft designers generally prevented armor from being added to aircraft, especially single-seat fighters, which relied on high performance and agility for their survival during air-to-air combat more than any other category of warplane. However, the availability of more powerful engines late in the war coupled with the increased need for ground attack and improved anti-aircraft weapons and tactics drove both Britain and Germany to investigate the need for armored fighters for ground attack and interception of other ground-attack aircraft.

Utilizing the most powerful rotary engine available during the war, Sopwith produced the Salamander armored fighter derived from their Snipe, the last Sopwith design introduced to combat during the war. The Salamander had good low-altitude speed and was placed in production, but only two were in France by the Armistice.

Both AEG and Fokker in Germany produced prototype armored fighters. The first, the AEG DJ.I triplane, was hobbled by its triplane configuration and a biplane development had to be built. The resulting DJ.I armored biplane fighter had reasonable low-altitude performance but used an experimental engine that was not yet in production, and the war ended before production was practical.

In the meantime, the innovative Fokker Company developed its V37 armored parasol monoplane fighter, similar in configuration to the last Fokker

Above: The AEG DJ.I biplane was developed from the earlier DJ.I triplane to improve its performance and flight characteristics. It was powered by the same experimental 195 hp Benz Bz.IIIb V-8 engine, which likely delayed its development. The biplane use the same armored fuselage as the triplane coupled to thick biplane wings unlike those on any other AEG design of the war. The thick wings enabled the design to dispense with vulnerable bracing wires.

Above: The Fokker V37 armored fighter prototype combined the configuration of the earlier Fokker E.V/D.VIII with a distinctive propeller spinner and a 195 hp Benz Bz.IIIb V-8 water-cooled engine. The entire forward fuselage of the Fokker V37 was armored with 2.5mm steel to protect the pilot, engine, and radiator from rifle-caliber ground fire.

fighter to reach the front, the E.V/D.VIII. The V37 was an interesting design with great potential, but also used an experimental V-8 engine that was not yet in production. Coupled with its late development, no production was possible before the Armistice, and in fact the war ended before a comprehensive test program could be completed. Perhaps the impending loss of the war sapped enthusiasm to develop the aircraft for production.

Germany started experimenting with armored fighters before Britain but could not bring one into production. Despite its later start, Britain actually brought its armored fighter into production, although too late to see combat. The result was the armored fighter never made an impact on the first air war and we can only speculate on what it might have achieved.

Seaplane Fighters

The first true land-based fighters, the Fokker *Eindeckers*, entered service in mid-1915 and soon proved themselves in combat. By the time the next generation of fighters appeared in combat, navies had started thinking about fighters based at sea. The technology and operating procedures to take land-based fighters to sea on aircraft carriers was still in the future, so the first naval fighters were of necessity seaplanes. In Northern Europe this meant floatplane derivatives of land-based fighters; in the warmer waters of Southern Europe flying boat fighters were also used.

Above: Sopwith Baby N2071 of No.229 Squadron, RAF, carried two Lewis guns and a 65-pound bomb. Powered by a 130 hp Clerget, it was used for short-range coastal patrols.

Left: Predecessor of the Sopwith Baby was the Sopwith Schneider seen here. A few, like this one, were supplied to the US Navy. Powered by a 100 hp Gnôme, later by a 110 hp Clerget, it was frequently used from shipboard to defend against Zeppelins and for reconnaissance.

Sopwith Baby N2071 of RNAS Yarmouth, 1918.

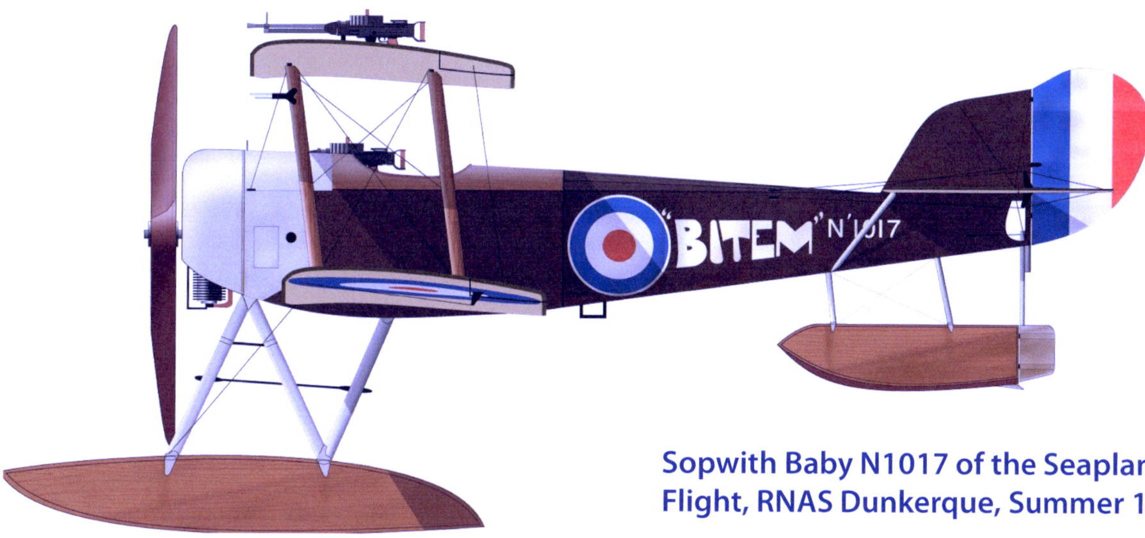

Sopwith Baby N1017 of the Seaplane Defence Flight, RNAS Dunkerque, Summer 1917.

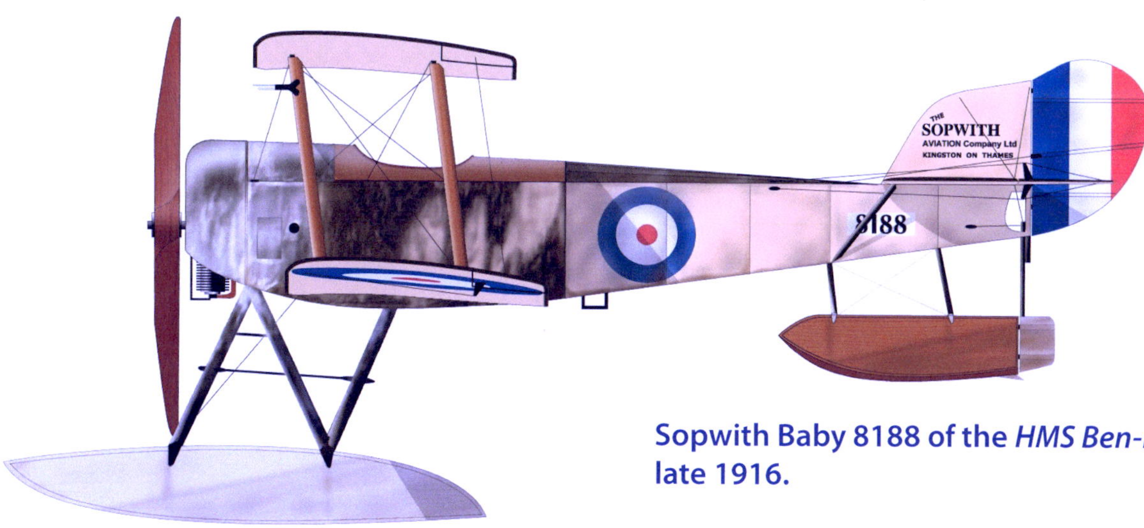

Sopwith Baby 8188 of the *HMS Ben-my-Chree*, late 1916.

Above & Below: The Spad 14 was the fastest floatplane fighter to see combat. A floatplane development of the unusual Spad 12, 47 were built. Like the Spad 12, the Spad 14 was armed with a single synchronized machine gun and a single-shot, hand-loaded 37mm cannon mounted in the 'V' of the Hispano-Suiza V-8 and firing through the propeller hub of the geared propeller. The Spad 14 served primarily on the Channel Front opposing German seaplanes.

Sopwith Pup N6453 flown on landing trials by Lcdr. E.H. Dunning aboard *HMS Furious* on August 7, 1917. Unfortunately, Dunning was killed on his next attempt while landing N6452.

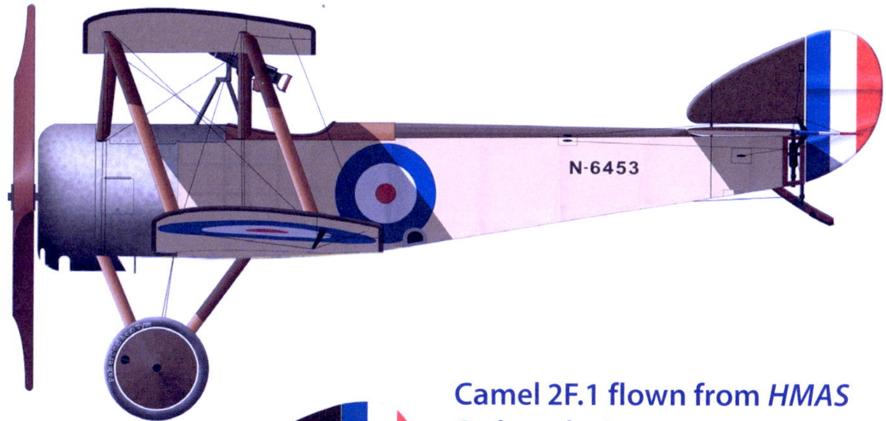

Camel 2F.1 flown from *HMAS Sydney* during summer 1918. The white-washed tires provided the pilot with a visual aid during take-offs from the small flying platform.

Camel 2F.1 flown by Capt. B.A. Smart from the *HMS Furious* on 19 July 1918 on the first carrier air strike in history. The target was the Zeppelin base at Tondern.

Hanriot HD.2 flown by George Moseley of the USN fighter squadron at Dunkerque NAS in 1918.

Hanriot HD.2 flown by Lt. Bamberry of the USN fighter squadron at Dunkerque NAS in 1918.

Above: The Hanriot HD.2 was a floatplane development of the HD.1 land-based fighter. The HD.1 was used by Belgium and Italy but not France; however, both the French and USN used the HD.2 on the Channel Front to escort Allied flying boats.

Above: Rumpler 6B1 Marine #751 is shown in camouflage colors after assignment to Zeebrugge. The 6B1 was a successful floatplane fighter developed from the Rumpler C.I reconnaissance two-seater. Flying from Zeebrugge, *Ltn.z.S.* Neimeyer used Rumpler 6B1 #751 to down a Short Seaplane on 31 Aug. 1916 and a Caudron G.4 on 7 Sept. 1916.

Above: *Oblt.zur See* Friedrich Christiansen in the Albatros W4 prototype at Zeebrugge. The W4 was a floatplane development of the Albatros D.I fighter. Christiansen went on to score 13 victories and was awarded the *Pour le Mérite.*

Above: This striking view of Marine #748, the first Brandenburg KDW prototype, shows the 'washed ou't ailerons. The KDW was a floatplane development of the D.I fighter used by Austria but not Germany. *Ltn*. Fritz Hammer used #748 to down a Russian Ilya Mouromets bomber in September 1916, one of only three downed by German aircraft in air combat. Although credited as a victory, the badly damaged bomber crash-landed at its base with three of it four crewmen wounded.

Left: Marine #946, the Brandenburg CC prototype, was armed with one gun and had a frontal radiator. It was the only flying boat fighter used by the German Navy. The German Navy did not favor flying boats for the cold waters of Northern Europe and the CC was soon replaced, but the CC served successfully with the Austro-Hungarian Navy in the warmer waters of the Adriatic.

Left: The German Navy's sole W18; its Marine Number 2138 is chalked on the side of the fuselage. The Brandenburg W18 was a faster, more powerful development of the CC and like the CC served successfully with the Austro-Hungarian Navy in the warmer waters of the Adriatic.

Left: The Ursinus U.1 floatplane fighter was unique in having retractable floats, a significant innovation for its time. To improve maneuverability the 150 hp Benz Bz.III engine was located on the center of gravity. Technical difficulties with the extended drive shaft to the propeller and the float retraction mechanism limited the design to a single prototype.

Albatros W4 Marine #911 after tactical markings were added. This fighter was flown by *Ltn.z.S.* Schulz, who downed three aircraft while flying it.

Marine #748 was the first KDW prototype. It was finished in stained wood for the fuselage and floats and clear-doped linen flying surfaces.

Rumpler 6B1 Marine #751 after camouflage was roughly applied at Zeebrugge. This is how the aircraft looked when flown on operations.

Above: W12s at their base at Zeebrugge. *Oblt.z.S.* Christiansen's #1183 is in the center.

Above: Germany's single-seat floatplane fighters were reasonably successful defending German naval air stations, but longer-range fighters of greater combat effectiveness were needed for offensive operations. The two-seat Brandenburg W12 fighter succeeded brilliantly and immediately became a problem for Allied flying boats over the North Sea. The innovative W12 set the pace for subsequent two-seat floatplane fighters and its monoplane derivatives served into the mid-1930s. The four-leaf clover on the fuselage of this W12 worked; this crew is happy to survive a forced landing on a grassy field. The W12 was a sturdy warplane and this aircraft turned out to be almost undamaged by its experience.

The Brandenburg CC prototype, Marine Number 946, in color shows its plain finish with stained wood fuselage and clear-doped linen flying surfaces.

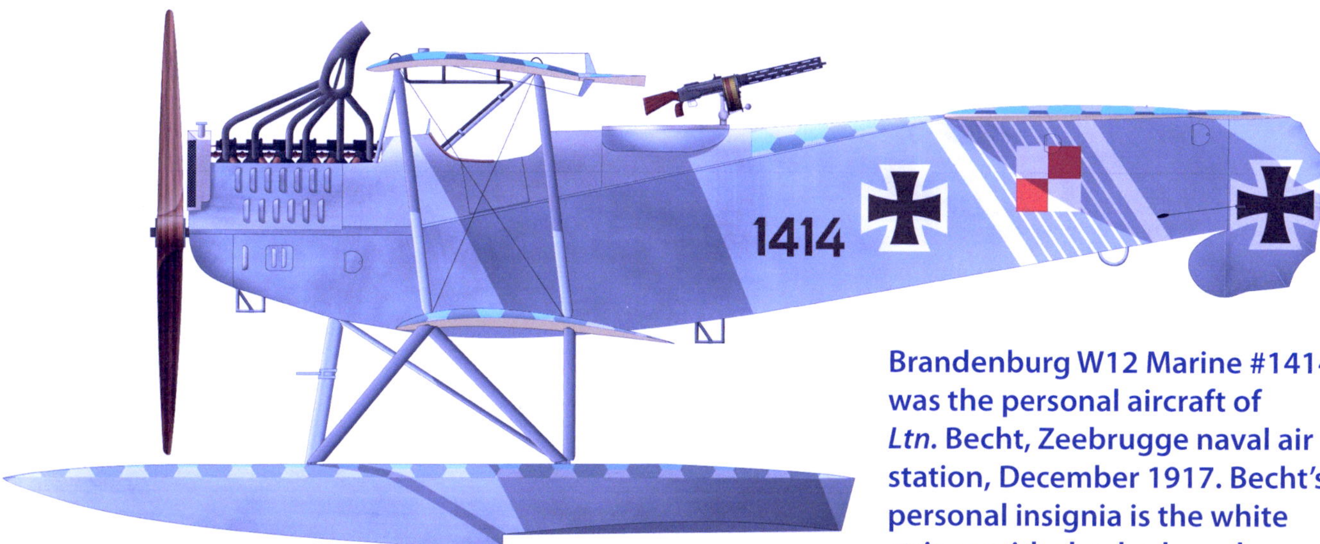

Brandenburg W12 Marine #1414 was the personal aircraft of *Ltn.* Becht, Zeebrugge naval air station, December 1917. Becht's personal insignia is the white stripes with checkerboard.

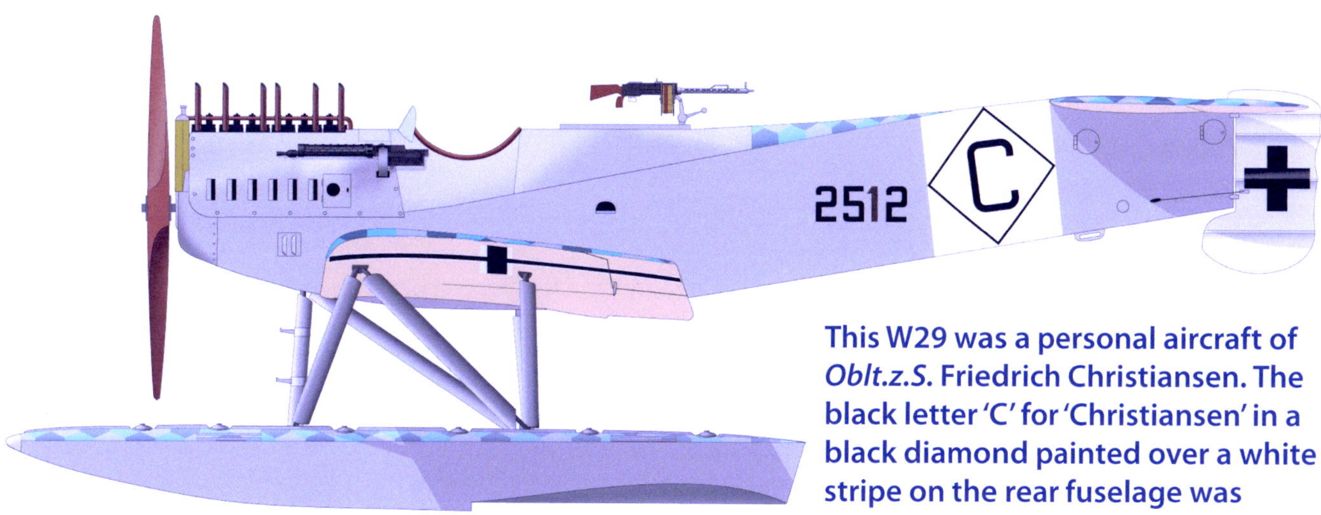

This W29 was a personal aircraft of *Oblt.z.S.* Friedrich Christiansen. The black letter 'C' for 'Christiansen' in a black diamond painted over a white stripe on the rear fuselage was Christiansen's personal insignia.

Above: The clean appearance and small spinner are shown in this view of Brandenburg W19 Marine Number 2207, the first production W19. The two-bay W19 was an enlarged, more powerful development of the single-bay W12 with more equipment and longer range.

Above: The W29 was a monoplane development of the W12 biplane for greater speed. Here W29 Marine #2532 from Norderney (indicated by the two identification stripes on the aft fuselage) on patrol over the North Sea. Like #2532, most W29s were category C3MG and had two fixed machine guns for the pilot. If a wireless transmitter and receiver were installed, the pilot had only one fixed gun to avoid over-loading the aircraft and reducing its performance. The Brandenburgs were at their most vulnerable when taking off and landing, and since they were based near the front there was always the possibility that they would be attacked by Allied land-based fighters. That was why the naval aircrews adopted the tactic of taking off and landing in formation, and maintaining a stepped Vee formation in flight with clear fields of fire for all the gunners. Without the handicap of floats British fighters, mostly Sopwith Camels, had the advantages of speed, climb, and maneuverability, so formation flying for mutual defense was a priority for the crews.

Above: The W33 was an enlarged development of the W29 powered by the 260 hp Mercedes D.IVa or similar engines and designed for greater range and endurance. Photographed at Warnemünde on 30 August 1918, these views of W33 Marine #2538 shows the very clean lines of the W33 despite its size. It was finished in the standard late-war naval camouflage. The W33 was the last Brandenburg design to reach combat in WWI but served in Finnland for many years after the war.

Above: The all-metal Junkers CLS.I was a floatplane development of the CL.I two-seat fighter. The CL.I was produced for the army and used postwar. The CLS.I was powered by a 200 hp Benz Bz.IV engine and was one of two German fifth-generation floatplane fighters. The CLS.I had a top speed of 180 km/h, which was somewhat faster than the W29. The corrugated metal skin was characteristic of early Junkers designs.

Above: The all-metal Zeppelin-Lindau (Dornier) CS.I was the second German fifth-generation floatplane fighter design to fly during the war. It is seen here during testing with the spinner removed. It still has the drag-producing ear radiators.

Below: The CS.I was intended to use a nose radiator as seen here. This installation produced less drag than the ear radiators and enabled the CS.I to attain greater speed. Stressed-skin all-metal construction was employed.

Oblt.z.S. Friedrich Christiansen, shown at right in a formal *Pour le Mérite* portrait reproduced as Sanke card 609, was the most famous and successful German naval floatplane ace. A very aggressive pilot, Christiansen took command of *Seeflugstation Flandern* I at Zeebrugge on the Belgian coast in September 1917 and is credited with at least 13 victories, including British airship *C27*, British submarine *C25* (which survived, although with its captain and a number of crew KIA), eight twin-engine flying boats, a Sopwith Pup, and two Short 184 floatplanes. He also set the Dutch schooner *Meeuw* on fire April 21, 1918.

THE NORTH SEA

Given the Royal Navy's great numerical superiority in battleships, Germany's only realistic chance to win the war at sea was submarine warfare – and those submarines had to transit the North Sea to attack British shipping. The Royal Navy maintained constant air and sea patrols to find and attack these U-boats, and German aircraft and Zeppelins patrolled the North Sea to support the U-boats against the Royal Navy. This set the stage for frequent clashes between British and German ships and aircraft. The 'Hornets of Zeebrugge' and floatplanes from nearby air stations featured prominently in these clashes and soon made the North Sea a dangerous place for Allied aircraft. In response, the Allies designed more powerful flying boats to survive these combats.

Brandenburg W12 #1183 was a personal aircraft of *Oblt.z.S.* Friedrich Christiansen. Christiansen took command of the Naval Air Station at Zeebrugge in September 1917; his personal insignia was the initial of his last name in a diamond on a white stripe as shown. Otherwise the aircraft was finished according to the naval directive of April 1917.

Seeflugstation Flandern I was the German naval airstation at Zeebrugge (Seebrugge) on Belgian's Flanders coast. The seaplanes from Zeebrugge operated over the North Sea, essentially the seaward extension of the Western Front. As the German seaplanes grew more technically capable they became more aggressive, especially once *Oblt.z.S.* Friedrich Christiansen took command. They became such a problem that they were nicknamed "The Hornets of Zeebrugge" and fought savage battles with the RNAS over the North Sea, attacking airships, flying boats, submarines, and coastal torpedo boats in some notable actions as shown in the photographs below.

Above Left: Friedrichshafen FF33L floatplanes from Zeebrugge on patrol over the North Sea sight a Dutch schooner.

Above: One of the Friedrichshafen FF33L floatplanes lands near the Dutch schooner to check it for contraband. German seaplanes captured or sunk a number of vessels like this that they found carrying contraband.

Left: The FF33L flown by *Oblt.z.S.* Friedrich Christiansen with his personal insignia on the rear fuselage is hanging from the crane at Zeebrugge.

Above: Brandenburg W29s operating with German warships in the North Sea.

Above: As illustrated here, longer-range W19s would frequently patrol with W12s and W29s, which sometimes waited on the surface until the W19 found a target and returned to lead the flight to it.

As the course of the maritime air war changed, so did the needs and procurement policies of the German Navy. Evolution of naval combat aircraft paralleled that of army aircraft to a large extent. Reconnaissance was the primary role of maritime aviation throughout the war, just as it was for the army. Initially the primitive reconnaissance seaplanes were unarmed. Airplanes of all types were evolving into larger, more powerful craft of longer range, greater speed, and the ability to carry increased loads, including armament. As air combat evolved from an oddity to routine, the reconnaissance seaplanes were armed.

Soon the Navy recognized the need for fighters to defend naval air stations against reconnaissance and bombing. Single-seat floatplane fighter prototypes were built, and all three types that entered production were modifications of Army aircraft already in use. These fighters were able to engage hostile seaplanes and two-seaters on favorable terms, but were inferior in performance to contemporary land planes due to the weight and drag of their floats. Their range and endurance were too short for offensive operations, and over-water navigation in a single-seat aircraft was also problematic.

These realities and the desire to conduct longer-range, offensive operations led the Navy to procure two-seat floatplane fighters. In addition to their longer range and endurance, the second crewman was an invaluable help with over-water navigation, and his flexible machine gun improved combat effectiveness. In addition, some of these aircraft carried a wireless transmitter/receiver operated by the observer, something not possible for a single-seat aircraft.

Brandenburg's innovative W12 performed better than its competitors and was produced in significant numbers for a naval aircraft. The great operational success of the two-seat W12 over the North Sea led to all subsequent production Brandenburg floatplane fighters being of similar configuration. Competing manufacturers were unable to design comparable aircraft until the innovative, all-metal Junkers CLS.I and Dornier Cs.I designs, which were too late for combat.

The North Sea was not only a seaward continuation of the trench lines on the Western Front, German submarines also had to transit

Above & Left: Two formations of Brandenburgs hunt over the North Sea. Brandenburgs often patrolled in formations of five aircraft; the trailing aircraft in the formation in the lower photo took the photo of the others.

Above: A W29 of Zeebrugge's 1-*C.Staffel* strafes Felixstowe No.4305 after it was forced down on July 31, 1918.

Above: Felixstowe No.4305 burns. The Brandenburgs earned their reputation as "The Hornets of Zeebrugge."

through it to their patrol areas, attracting British airships and flying boats on anti-submarine operations. With the appearance of the Brandenburg W12 the Germans were now able to effectively counter those operations.

For even greater range and endurance than the W12, a larger, more powerful aircraft was needed, and Brandenburg's W19, basically an enlarged W12, was produced in response to that need.

Although the W12 was faster than the big Felixstowe flying boats that were its frequent antagonists, its closing speed was not as great as desired. A fleeing Felixstowe could sometimes evade a W12 long enough that the W12 might have to turn back due to low fuel state before completing the interception. So, for greater speed, the biplane W12 was developed into the monoplane W29. The reduced drag made the W29 faster, which enabled it to intercept the Felixstowes much more quickly to ensure they did not escape.

Just as there was a need for a larger, longer-range development of the W12, the same was true of the W29, and the W33 combined the W29 monoplane's configuration with the larger size, greater fuel capacity, and increased range and endurance of the W19. The W29 and W33 were difficult to distinguish during combat, and there were few of them, resulting in the first type to arrive at the front, the W29, getting all the recognition.

The W33 was to have been followed by the further enlarged W34, but the W34 did not enter service before the armistice. Postwar the similar W37 design was built in Germany and assembled in Sweden as the Caspar S.1. Of 305 Brandenburg monoplanes ordered, 209 were delivered during the war.

The W29 and W33 monoplane floatplane fighters were robust, maneuverable, and fast for their time, and served successfully for many years after the war in demanding Nordic environments, being built under license by Denmark, Norway, and Finnland.

An interesting sidelight to the story of German seaplane fighters is the great variety of types produced considering their comparatively limited production quantities.

Above: British non-rigid airship C27 goes down in flames after an attack by Brandenburg W12s led by *Oblt.z.S.* Christiansen over the North Sea on 11 December, 1917.

Above: A W29 attacks British Submarine *C25* that was caught on the surface on 6 July 1918. Armed only with machine guns, the Brandenburgs penetrated the pressure hull, preventing the submarine from safely submerging.

Above: British submarine *C25* under attack by Brandenburg W29s on 6 July 1918. Their gunfire killed the captain and five crewmen and damaged the submarine so badly it was forced to return to port for repairs. For this action *Oblt.z.S.* Friedrich Christainsen was credited for a victory over *C25*.

Flying boat fighters from both the Austro-Hungarian Empire and Italy were active over the Adriatic during the last part of the Great War, frequently in combat with one another and with land planes. In addition to two-seat flying boats, Austria added the Brandenburg CC single-seat flying boat fighter, later replaced by the faster and more powerful Brandenburg W18. At first Italy responded with two-seat flying boats, and later with the single-seat Macchi M.5 flying boat fighter, the fastest and probably the best flying boat fighter of the war. The M5's performance was so good that the Austro-Hungarian Navy was forced to respond with Phönix D.I land-plane fighters.

By 1918 time America had entered the war, and US Naval Aviators were trained in Italy and flew combat missions there in Macchi M.5 fighters. This lead to one of the eight Medal of Honor awards to an American aviator during the war. On August 21, 1918, during a bombing raid on the Austro-Hungarian port of Pola, Landsman for Quartermaster Charles Hammann landed his M.5 at sea to rescue his squadron mate Ensigh George Ludlow, who had just been shot down by Austrian naval ace Friedrich Lang flying a Phönix D.I. Hammann performed his daring rescue despite the fact that the air combat was still in progress. For his heroism, Hammann was awarded the Medal of Honor, the only such award to a US Naval Aviator during the war.

Above: *Fregattenleutnant* Stephan Wollemann with his Phönix D.I A-118, shot down Ensign Charles Ludlow for his only confirmed victory of the war.

Above: Phönix D.I A-98 at Altura, Winter 1917–1918.

Below: The Italian Macchi M5 was the fastest flying boat fighter of the war and had good maneuverability. It was also flown by a small number of US Naval Aviators in Italy.

Right: Landsman for Quartermaster Charles Hammann was awarded the Medal of Honor for his actions on August 21, 1918 while flying a Macchi M.5 flying boat fighter from the Italian naval air station of Porto Corsini.

Above: With five victories, *Fregattenleutnant* Friedrich Lang was the only other Austro-Hungarian naval ace beside Baron Gottfried Banfield. Flying Phönix D.I A-117, Lang was one of the pilots who engaged Ludlow and Hammann.

Below: Ensign George Ludlow was flying this Macchi M.5 fighter *Mutt 2nd* during the attack on Pola on August 21, 1918 when he was shot down by *Fregattenleutnant* Friedrich Lang.

Gottfried *Freiherr* von Banfield was the complete package; nobleman, movie-star looks, Austria's leading naval ace of the war, and the last surviving holder of the Military Order of Maria Theresa, the highest award of the former Austro-Hungarian Empire, that brought with it elevation to the nobility. He was also a true gentlemen. Known as the "Eagle of Trieste", he may also be the only ace who flew a flying boat to score five or more victories.

Banfield is credited with nine confirmed and 11 unconfirmed victories. Most of his combats were over the northern Adriatic and his victims came down at sea, accounting for his high proportion of unconfirmed victories.

Banfield had a long and successful postwar career, resettling in Trieste, now an Italian city, in 1926 after a time in England. There Banfield eventually took over the Diodato Tripovich & Co. Trieste Shipping Company from his father-in-law and even won a local tennis championship in 1927. Banfield became a local celebrity, usually being called "Our Baron" by his former adversaries. Honored by friend and foe alike, Banfield passed away in Trieste in 1986 at the age of 96!

Right: In September 1917 a Hungarian magazine published this picture of Gottfried Banfield on the cover with the caption "Our Richthofen".

Right & Below: Among the flying boats Gottfried Banfield flew on operations is this one of a kind Oeffag H serial A.11. It is shown here at the Austro-Hungarian Trieste Naval Air Station in the summer of 1917. Such was Banfield's reputation by this time that Josef Mickl designed this fighter especially for him. Banfield had it painted blue overall and it was subsequently known as the "Bluebird". Banfield liked the Oeffag H and urged its mass production, but that was not undertaken. It is on its beaching dolly and could be rolled directly to the water for takeoff from here.

Right: The speedy SVA 5 fighter/reconnaissance plane was adapted to floats as an interceptor to defend Italian naval air stations.

Macchi M.5 '6' flown by the 260a *Squadriglia*, Venice.

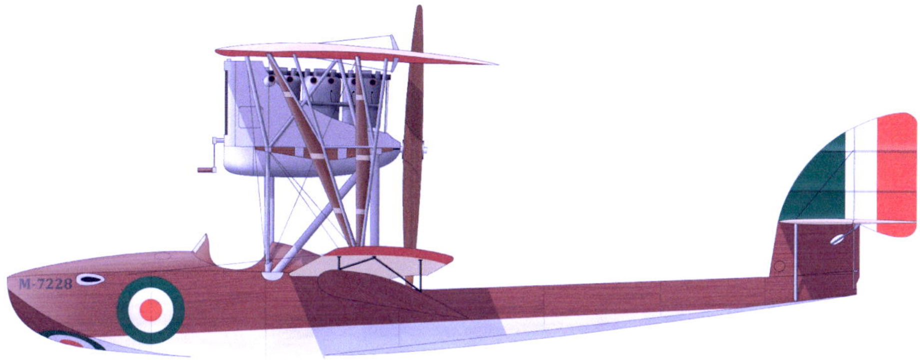

Macchi M.5 flown by Landsman for Quartermaster Charles Hammann during his Medal of Honor mission on 21 August 1918.

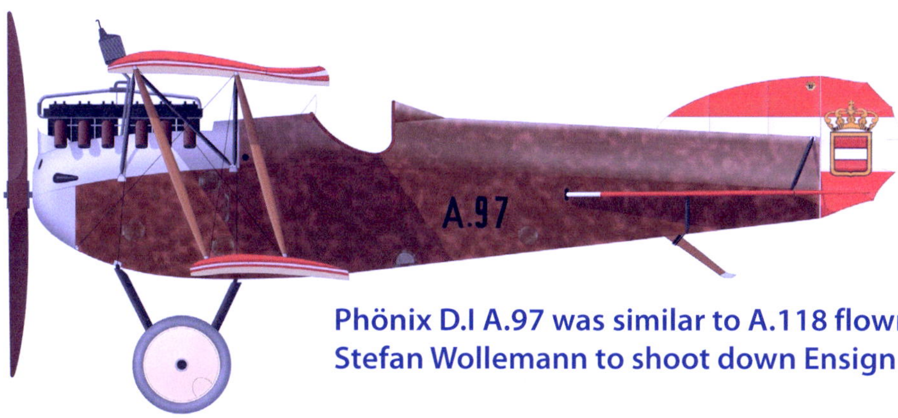

Phönix D.I A.97 was similar to A.118 flown by *Fregattenleutnant* Stefan Wollemann to shoot down Ensign Charles Ludlow.

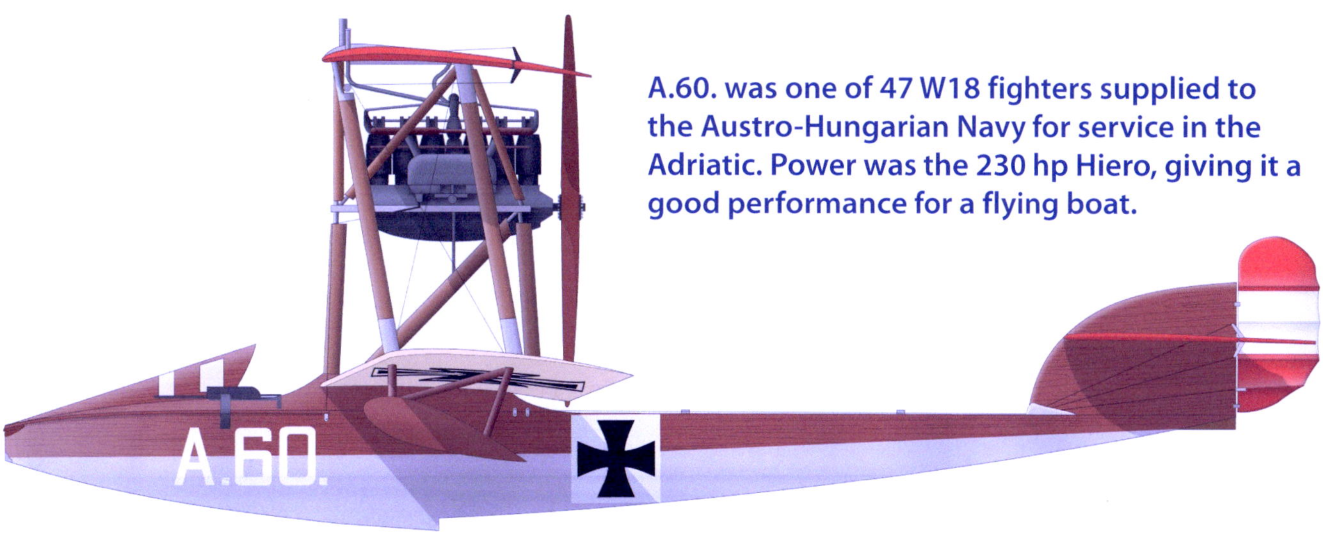

A.60. was one of 47 W18 fighters supplied to the Austro-Hungarian Navy for service in the Adriatic. Power was the 230 hp Hiero, giving it a good performance for a flying boat.

Above: The Port Victoria P.V.2 was originally designed as an anti-Zeppelin fighter with a Davis gun. When the Davis gun was abandoned, it was modified into a conventional fighter as the P.V.2bis seen here, which remained a prototype. It was powered by a 100 hp

Above & Below: The Port Victoria P.V.4 two-seat fighter remained a prototype. Designed for an experimental 150 hp engine that did not materialize, a 110 hp Clerget was fitted. The gunner sat in front with a Lewis gun. It did not fly until 1917; by that time pushers were obsolete.

Above & Right: The Port Victoria P.V.5a single-seat fighter-bomber remained a prototype. The P.V.5 (above) had a 150 hp Hispano-Suiza and the P.V.5a (right) had a 200 hp Hispano-Suiza. Despite good performance and maneuverability, by the time they were tested it was decided they were no longer needed and the designs were abandoned.

Most of the Royal Navy's aerial missions demanded long-range reconnaissance, and seaplane fighters remained low priority, with the Sopwith Schneider and Baby fulfilling the requirement throughout the war. Better fighters were designed but not placed in production.

Left: The Port Victoria P.V.9 had a new 150 hp Bentley B.R.1 rotary engine whose teething problems delayed its development. Despite that it had good performance and maneuverability and was considered the best single-seat seaplane fighter built to that time. However, again there was no service requirement and the design was abandoned.

Shipboard Fighters

Through the end of WWI the Royal Navy had been the world's dominant navy for more than a century. Thus it was natural that the Royal Navy would be interested in aircraft. All combatants developed flying boats and floatplanes, but it was evident that these could not match the performance of land planes due to their heavy, drag-producing floats or hulls. Consequently, the Royal Navy took the initiative to operate land planes from ships, resulting in the world's first aircraft carrier and carrier operations.

First, suitable aircraft with low stalling speeds like the Sopwith Pup were tested. This research lead to use of the more powerful Sopwith Camel from both shipboard and sleds towed by destroyers. Camels were used offensively in the first carrier air strike in history in August 1918, and in the same month were successful when flown defensively on anti-Zeppelin operations from the towed sleds.

Numerous combatants operated seaplanes from ships during WWI, but only the Royal Navy was able to operate land planes from ships, a development widely used today by larger navies.

Right: Lcdr. E.H. Dunning landing Sopwith Pup N6453 on the flight deck of *HMS Furious* on 7 August 1917. He switched to N6452 for his next attempt. Unfortunately, he stalled on landing, went overboard, and was drowned.

Left: Sopwith Camel 2F1 taking off from *HMS Furious*, the world's first aircraft carrier. The 2F1 was a special version of the Camel designed for shipboard use. The fuselage folded to conserve space, the wing span was reduced by a foot, and the armament was modified to one Vickers and a Lewis over the wing, the Lewis being able to fire upward at Zeppelins, which were used for maritime reconnaissance over the North Sea.

Above: Sopwith 2F1 Camels on *HMS Furious* en route to the attack on the Zeppelin sheds at Tondern, the world's first carrier air strike. These seven Camels launched on July 19, 1918 and bombed and destroyed Zeppelins *L.54* and *L.60* in their sheds; one Camel and its pilot were lost. As a result of this successful raid, the Germans abandoned Tondern as an active Zeppelin based and strengthened their defenses at other Zeppelin bases.

Below: The Royal Navy needed aircraft at sea to intercept Zeppelins on their reconnaissance missions, and seaplanes generally did not have sufficient performance to successfully intercept the new Zeppelins. One solution has to launch high performance land-plane fighters at sea, and Sopwith 2F1 Camels were carried on lighters towed behind Destroyers for that purpose. On August 11, 1918, Lt. Culley flew a Camel off a lighter to intercept Zeppelin *L.53*. Culley climbed to more than 18,000 feet and reached 300 feet below the Zeppelin. Unable to climb higher, Culley pulled the Camel's nose up and fired; *L.53* caught fire and fell in flames, the last Zeppelin to be shot down during the war.

Above: The Beardmore W.B.IV was a prototype single-seat fighter designed to operate from ships. The 200 hp Hispano-Suiza engine was mounted over the lower wing and drove the propeller via an extension shaft. The pilot sat in front of the wing for excellent vision. The hull was designed to float if the aircraft were forced down at sea, and the pilot could jettison the undercarriage if forced to ditch. Armament was a single fixed, synchronized machine gun. Only one prototype was built.

Left: The Mann, Egerton H.1 was designed as a shipboard fighter; power was a 200 hp Hispano-Suiza. Note the floatation gear under the wings and fuselage in case of emergency landing at sea. The wings could be folded for storage aboard ship. Under test the H.1 flew well but failed the floatation test.

Right: The Mann, Egerton H.2 shipboard fighter; also powered by a 200 hp Hispano-Suiza, dispensed with the external floatation gear in favor of inflatable air bags. The H.2 was tested but not produced, and only a single example of each type was built.

Fighter Engines

Pilots know that the engine is the heart of the aircraft, so a brief discussion of WWI fighter engines is in order.

One of the first things one notices about WWI fighters is that most German fighters had six-cylinder, water-cooled, inline engines; the notable exceptions among production aircraft were the Fokker *Eindeckers* and Triplane, which had rotary engines. In contrast, most production Allied fighters used rotary engines until the advent of the Hispano-Suiza V-8. After the 'Hisso' became available it was used to an ever greater extent that was limited by its availability. What drove this difference?

To fly, airplanes must have a lot of power compared to their weight, so engines that generate high power for their weight are essential. And the key advantage of rotary engines was their high power for their weight. Part of that advantage was being air cooled and not needing a heavy radiator and coolant, which gave them the secondary advantage of no vulnerable cooling system.

However, in all other respects the rotary was inferior to powerful water-cooled engines. Rotaries used much more fuel and oil than water-cooled engines of similar power, were far less reliable, and many did not have a flexible way to throttle the engine. Moreover, Castor oil was by far the best lubricant for rotaries.

All these problems led to the disappearance of the rotary soon after the war ended; new engine

Above: The Sopwith Camel was the classic rotary-powered British fighter. The Camel was an excellent dog-fighter but could not match the SE.5a in performance, especially at higher altitudes.

Below: The SE.5a, this one flown by British ace James McCudden, 57 victories, was the classic Hispano-Suiza-powered British fighter.

Above & Above Right: From the time the first Albatros fighters arrived at the front powered by the 160 hp Mercedes, Germany relied on the six-cylinder water-cooled engine for most fighters. Above left an Albatros D.Va on display at the Australian War Museum shows off its engine. The radiator is mounted in the root of the upper right wing and is just visible. Germany's Austro-Hungarian allies faced the same issues; above right is Berg D.I 101.37 at the Vienna Technical Museum with its six-cylinder, water-cooled engine. An additional advantage of water-cooled engines was they retained their power better than rotary engines as altitude increased.

Above: The classic, rotary-powered Nieuport fighter series was critical to Allied pilots well after the arrival of the Hispano-Suiza V-8 engine due to the difficulty in producing it in sufficient quantity. Here Belgian ace Jan Olieslagers is photographed in front of his Nieuport 11.

development focused primarily on water-cooled engines and stationary radial engines. The rotary reached the limit of its development potential with the 230 hp Bentley B.R.2 (used in the Sopwith Snipe) and the 205 hp Siemens-Halske Sh.IIIa used in the SSW D.III and D.IV, the Pfalz D.VIII, and some prototypes. By the time these engines appeared at the front, prototype Hispano-Suiza V-8 engines were developing 300 hp, and designers looked to the V-8 and V-12 configurations for the greater power necessary for the new aircraft being built.

The early fighters of both sides used rotary engines due to their high power to weight ratios. Once the 160 hp Mercedes D.III inline six became available to the Germans and was introduced to fighter combat by the Albatros fighters, nearly all German fighters used the Mercedes or its later derivative, the BMW. The Mercedes was powerful, far more reliable than rotary engines, and much more economical of fuel and oil, essential benefits to Germany which had very limited oil resources. Furthermore, Germany had very limited stocks of Castor oil, and once those were exhausted the substitutes tried did not work well, further reducing rotary engine reliability.

Above: Availability of the Hispano-Suiza V-8 engine enabled production of the Spad 7 and its developments. The Spad was much faster and more robust then the Nieuport series and was superior to German fighters until the exceptional Fokker D.VII arrived in April/May 1918. Here Belgian ace Edmond Thieffry poses with his Spad 7.

The Allies, with ample stocks of Castor oil, continued to use rotary engines in many fighters until the end of the war. The availability of the Hispano-Suiza V-8 gave some of the same advantages as the Mercedes; eventually the 'Hisso' gave usefully more power but it was never as reliable. The Hisso V-8 was a breakthrough for Allied fighter design and its adoption was limited primarily by its production. Germany's belated recognition of the value of the V-8 configuration and inadequate development priority ensured that no German aircraft flew in combat powered by a V-8 engine. German V-8 engines did reach production, but too late to power a combat aircraft over the front. Worse yet, Germany aimed for an engine competitive with the Hispano-Suiza engines being captured in 1917; by late 1918 more powerful Hissos were already in production.

German development of more powerful rotary engines also lagged Allied progress. Until well into 1918, production German rotaries were limited to 110–120 hp with the exception of the unreliable and heavy two-row 160 hp Oberursel U.III. The innovative Siemens-Halske Sh.III counter-rotary engine that typically developed 205 hp at low propeller RPM was the only advanced rotary to power German fighters into combat. Other advanced German rotary engines were still in development or limited to pilot production until the war's end and did not reach the front.

Fighter Genesis in Retrospect

Although biplane fighters dominated air combat during WWI due to their comparatively light weight and low drag, the very first purpose-designed single-seat fighters were the Fokker monoplanes. The Fokkers had reasonably good performance and maneuverability for their time, but their key advantage was their effective armament, a machine gun fixed to fire forward along their line of flight and synchronized to shoot through the propeller arc.

The Fokker monoplanes revolutionized air combat and their early pilots became the world's first aces, with a major impact not only on the air war but on popular culture. During their many months of dominance over the Western Front the vulnerability of Allied reconnaissance airplanes became a scandal and their crews became known as 'Fokker Fodder'.

The French Nieuport 11 and British DH.2 biplane fighters, both powered by rotary engines similar to those in the Fokker and Pfalz *Eindeckers*, arrived at the front in early 1916 and soon demonstrated superior climb and maneuverability compared to the *Eindeckers*, which were gradually replaced by superior German biplane fighters as those became available. The *Eindeckers* thus started the first technology race between opposing aircraft.

There is no mystery about the biplane quickly superseding the monoplane at the time. Given the primitive materials and structural design techniques then prevalent, the biplane matched the monoplane in weight and drag while being more robust and maneuverable. Its greater strength also enabled the effective use of more powerful engines. The extensive wire bracing monoplanes needed was heavy and created excessive drag, and the resulting structure was not robust enough as fighter combat developed into high-G maneuvering contests.

Although the Allies also had monoplane fighters, especially those built by Morane-Saulnier, these lost favor for the same reasons the *Eindeckers* disappeared. Despite some good designs, Allied pilots and commanders were skeptical of the monoplane's strength and standardized on biplanes.

A key problem for Germany was stagnation of aero-engine design. Once the excellent 160 hp Mercedes D.III was established in mass production, followed by the dramatic impact of the Albatros fighters, *Idflieg* was content to mass-produce the Albatros fighters and the Mercedes that powered them. Fearing disruption of engine production by introduction of a new type, *Idflieg* failed to push engine development with enough vigor, with the result that the next generation of fighter designs had to use existing engines. This was a serious mistake; the engine is the heart of an aircraft and determines its performance potential. However, instead of developing new engines, which takes extensive time and resources that were limited in wartime Germany, *Idflieg* instead focused on advanced structures and aerodynamics to achieve better performance.

In contrast, the resource-rich Allies pursued engine development to increase power for better performance and relied on existing structural methods and materials to facilitate mass production.

The Allies fielded aero-engines of many configurations, not only rotary engines but water-cooled radial engines and both air and water-cooled V-8 and V-12 engines. Shockingly from the German standpoint, no WWI German warplane flew an operational mission powered by a German V-8 or V-12 engine. Even after *Idflieg* directed the German aero-engine manufacturers to develop light-weight V-8 engines for fighters, the companies were unable to deliver production V-8 engines in quantity before the armistice. This failure was despite prolonged development and the ready availability of captured Hispano-Suiza V-8 engines for reference. Similarly, the excellent line of Rolls-Royce V-12 engines were never matched in Germany. And while the

Left: Oswald Boelcke and his ground crew pose in front of Fokker E.IV 123/15, the second E.IV built. The E.IV was the last of the early Fokker monoplane fighters; new biplanes had superior performance and maneuverability. Boelcke is the father of German fighter aviation; the Red Baron, one of his students, said of him, "I am just a combat pilot but Boelcke, he is a hero."

American Liberty Plane was criticized for producing a 1916 airframe during 1918, the V-12 Liberty engine that powered it was very successful and Germany had nothing comparable.

One result of Germany's distress over her fighter situation in mid-1917 was the Triplane Craze that afflicted *Idflieg*, and by extension the German aviation industry, for so much of 1917.

The biplane would have remained ascendant except for the work of two famous designers, Hugo Junkers and Anthony Fokker. Junkers was a scientist interested in low drag, all-metal aircraft and did the pioneering work on cantilever monoplane wings based on a relatively thick airfoil. Junkers and Fokker were forced to collaborate by *Idflieg*, who wanted Fokker to help solve the manufacturing problems being experienced by Junkers in building all-metal airframes. Fokker built his airplane fuselages out of welded steel tubing, which gave *Idflieg* the mistaken idea he could be of help. But Junkers was using duralumin for the entire airframe, a much more difficult proposition, plus the men's personalities were totally incompatible, so nothing came of this forced association for Junkers.

The impact on Fokker was more fortuitous; he became interested in the thick, cantilever wing designs he saw at Junkers and worked with a major German plywood manufacturer to see if cantilever wings could be built of wood. The project was successful and was applied to biplane test aircraft and then the famous Fokker Triplane; its interplane struts were unnecessary but were added to reassure its pilots. The same technology was applied to biplanes and monoplanes, eventually resulting in the Fokker E.V/D.VIII that is famous despite its very limited combat experience.

Junkers and Fokker led the way to the cantilever monoplane fighter and both designed aircraft that reached operational service in small numbers before the armistice. There is no doubt that by late 1918 the monoplane configuration was almost as popular in Germany as the triplane had been in the middle of 1917. But unlike the triplane craze, which was based on emotion, the monoplane's appeal was based on performance – made possible by advanced structural designs. The Daimler L11 and SSW D.VI monoplanes of late 1918 had much better performance than contemporary operational Allied fighters.

The secret of a successful monoplane design was the wing structure and aerodynamics, and the thick airfoil was key to both. The thick airfoil enabled a robust structure at reasonable weight while eliminating drag-producing external bracing struts and wires, and also gave docile stall characteristics and high lift. Junkers and Fokker understood this; other designers of the time did not.

The successful German monoplane fighters of 1918, now mostly forgotten, previewed the future of aircraft design for the next three decades of the 20th Century.

Different Design Philosophies

A good example of the difference in the German and Allied design approaches is the contrast between the last fighters introduced into combat before the armistice. Germany introduced the Fokker D.VIII, a parasol monoplane that was the minimum practical fighter. It had an innovative, cantilever wooden wing and welded steel tube fuselage and tail covered by fabric. These materials were readily available even in Germany. Reaching combat in August 1918, it was powered by a rotary engine nearly identical in power to that used in the Fokker E.III in the summer of 1915! The D.VIII thus combined a 1915 vintage engine – dictated by resource shortages and complacency – with 1918 structural and aerodynamic designs.

In contrast the Sopwith Snipe, a larger, more powerful derivative of the 1916 Camel, was powered by a new 230 hp Bentley B.R.2 rotary engine, the most powerful rotary of the war. The Snipe combined a 1918 engine with 1916 structural design and aerodynamics. Snipe development was prolonged in a futile attempt to match the handling characteristics of the new Fokker D.VII. While the Snipe offered good performance due to its powerful new engine, its thin wing precluded it from matching the lift and benign stall qualities of the thick-wing Fokker. Furthermore, the additional weight and drag of the Snipe's two-bay wing limited its performance advantage over the older Sopwith Camel. Interestingly, the D.VIII and Snipe were competitive in speed, with the lighter D.VIII having superior climb despite having less than half the power (110 hp vs. 230 hp) of the Snipe.

Evolution of Tactics

Fighter tactics evolved with combat experience and aviation technology. First were the lone-wolf hunters in their early fighters who stalked the enemy. They were followed by squadrons of fighters that flew together and gradually developed team tactics. As fighters became more numerous, the size of formations continued to increase and squadrons were combined into wings to seize air superiority with a combination of training, tactics, technology, and numbers. With numerical superiority the Allies were first to practice this approach and the Germans were forced to respond with their *JG*1 (*Jagdgeschwader* 1), Germany's first fighter wing, the famous 'Flying Circus' commanded by Manfred von Richthofen.

Left: This unarmed Fokker D.VIII, undergoing post-war performance tests at McCook field in the US, represented the minimum practical fighter aircraft for the time. Its sophisticated wing and simple steel-tube fuselage were combined with a low-power engine, exemplifying the German focus on structures and aerodynamics for performance.

Right: The Sopwith Snipe was the last new British fighter to arrive at the front. The Snipe combined a powerful, 230 hp rotary engine introduced in 1918 with structural and aerodynamic technology essentially the same as 1916 Sopwith designs, exemplifying the Allied focus on more powerful engines for improved performance. (Alex Mitchell, The Vintage Aviator Collection)

Seaplane fighters were never numerous and were used in small formations. Night fighters necesarily flew independently, and armored fighters never reached the front. Convoy escort fighters operated with modest success, although the Caudron R.11 was the exception and was very successful in combat.

More important operationally were the two-seat fighters. With the 'sting in the tail' and, for seaplanes, much better navigation offered by the back-seater, two-seat fighters grew in importance until the Allies planned to field more two-seat fighters than single-seat fighters in 1919. Had the war lasted into 1919 the prevalence of two-seat fightes would have changed the fighter pilot legend, although to what degree is unknown.

Left: The Zeppelin D.I, with its all-metal stressed-skin construction, previewed the combat aircraft of the next great war to come, although most of those aircraft would be cantilever monoplanes instead of cantilever biplanes like the D.I. Despite its biplane configuration, the Zeppelin D.I had the most advanced structure of any WW1 fighter.

Fighter Pilots & Popular Culture

Above & Above Right: The film posters for the 1930 (above left) and 1938 (above right) films of *The Dawn Patrol*.

Without question the mythos of the fighter pilot created in WWI made a tremendous impact on popular culture. Surprisingly, a few important aces such as Manfred von Richthofen (the Red Baron) and James McCudden wrote their autobiographies during the war, and they were followed by many other aviators after the war. And it is not a surprise that Hollywood both rode the wave and fueled it with lavish productions in the late 1920s and 1930s.

This was the start of the trend but hardly the end; after another huge boost during and after WWII the trend continues to the present day. As aviation technology advanced, so its importance to combat success grew more critical. And if the aircraft today are far more reliable and effective than the primitive birds of WWI, the hazards of going to war in a cockpit continue to be extreme and require courage, skill, and ability.

The story of combat aviation in general, and fighter aviation in particular, is both dramatic and intensely human. It thus brings new life to the old stories of knightly combat and heroic and chivalrous conduct that people have told for generations – and this time with more truth in them!

One of the interesting differences between the myths of knights in shining armor and the stories of combat pilots, especially fighter pilots, is the difference in their regard for authority. If knights often became nobles, at least in the stories, and identified with the political leadership of their era, that is much less true of pilots. In the modern era the terrible reality of war has become far too well-known for many of the old illusions to survive, and the failures of leadership and vision that led to The Great War and following conflicts have destroyed much of the authority of the authority figures. A large part of the reputation of fighters pilots concerns their distain for authority. In fact, this popular opinion

Above: The 1938 version of *The Dawn Patrol* used an all-British cast, except for Australian Errol Flynn, and played very well together both in the light-hearted early scenes and grim later scenes.

has a significant basis in fact that can be traced back to the very beginning of fighter aviation in 1915. Famous aces such as Jean Navarre, the first French 'ace of aces', were notorious for their deep and abiding lack of regard for authority, a situation correlated

Above: The chivalry of fighter pilots has been exaggerated, but there was some. Schoenfelder of *Jasta* 7 has his photo taken in front of his Fokker D.VII with the RAF pilot he has just shot down. At least his chagrined captive is alive!

Above: French ace George Guynemer, who scored 53 victories before his death in action, mounted a camera on his fighter and took some of the world's first gun-camera photos. Here he pursues an Albatros D.III.

to pilots' desire to defy other traditions to fly fragile machines through the air. Anyone familiar with the armed services will be aware of significant personality differences between the different branches.

Another aspect of the 'knights of the air' was their supposed chivalry. Perhaps surprisingly amidst the horror and mass slaughter of the war, there actually was some. When the Red Baron was killed in combat, his British adversaries gave him a funeral will full honors (below) and notified the Germans they had done so. Many German pilots hosted their defeated foes for dinner at the *Jasta* before sending them on their way as PoWs. French ace George Guynemer saw Ernst Udet struggling with jammed guns and waved and dove off, sparing Udet's life.

Above: An Australian honor guard fires a salute volley during Richthofen's funeral ceremony in Bertangles cemetery on 22 April, 1918

Appendix 1: Leading Aces and their Scores

The victory totals for the aces listed here are their officially confirmed scores. Discussion of the accuracy of these totals and the differing criteria for confirmation of a victory by different nations is beyond the scope of this work. The symbol '†' after the ace's name indicates the ace was killed during the war, either in action or in a flying accident.

France

Ace	Score
Capt. René Fonck	75
Capt. Georges Guynemer †	53
Lt. Charles Nungesser	43
Capt. Georges Madon	41
Lt. Maurice Boyau †	35
Lt. Michel Coiffard †	34
Lt. Jean Bourjade	28
Capt. Armand Pinsard	27
S-Lt. René Dorme †	23
Lt. Gabriel Guérin †	23
S-Lt. Claude Haegelen	22
S-Lt. Pierre Marinovitch	21
Capt. Alfred Heurtaux	21
Capt. Albert Deullin	20
René Fonck was the highest-scoring Allied ace.	

Germany

Ace	Score
Rittm. Manfred von Richthofen †	80
Oblt. Ernst Udet	62
Oblt. Eric Lowenhardt †	53
Lt. Josef Jacobs	48
Lt. Werner Voss †	48
Lt. Fritz Rumey †	45
Hptm. Rudolf Berthold	44
Lt. Paul Baümer	43
Hptm. Oswald Boelcke †	40
Lt. Franz Büchner	40
Lt. Lothar von Richthofen	40
Lt. Heinrich Gontermann †	39
Oblt. Carl Menckhoff	39
Manfred von Richthofen, known as The Red Baron, was the highest-scoring ace of the war.	

Italy

Ace	Score
Maj. Francesco Baracca †	34
Ten. Silvio Scaroni	26
Ten. Col. Pier Piccio	24
Ten. Flavio Baracchini	21
Capt. Fulco Ruffo di Calabria	20
Ten. Ferruccio Ranza	17
Sgt. Marziale Cerutti	17

Austro-Hungarian Empire

Ace	Score
Hptm. Godwin Brumowski	35
Offz. Julius Arigi	32
Oblt. Benno Fiala	28
Oblt. Frank Linke-Crawford †	27
Lt. Josef Kiss †	19
Lt. Franz Gräser †	18
Fldwb. Eugen Bönsch	16
Sfdwb. Stefan Fejes	16
Oblt. Ernst Strohschneider †	15

Belgium

Ace	Score
Lt. Willy Coppens	37
Lt. Andre Demeulemeester	11
S-Lt. Edmond Thieffry	10
Cmdt. Fernand Jacquet	7
Lt. Jan Olieslagers	6
Note: 35 of Coppen's 37 victories were balloons.	

Russia

Ace	Score
Maj. Aleksandr Kosakov †	20
Ens. Vasili Yanchenko	16
Capt. Paul d'Argueff	15
Lt. Ivan Smirnov	11

United Kingdom	
Ace	**Score**
Lt. Col. William Bishop	73
Lt. Col. Raymond Collishaw	62
Maj. Edward Mannock †	61
Maj. James McCudden †	57
Capt. Anthony Beauchamp Proctor	54
Maj. Donald MacLaren	54
Maj. William Barker	50
Capt. Robert Little †	47
Capt. George McElroy †	46
Capt. Albert Ball †	44
Maj. Tom Hazell	43
Capt. Philip Fullard	40

USA*	
Ace	**Score**
Capt. E. Rickenbacker	26
Lt. Frank Luke †	18
Maj. Raoul Lufberry †	16
Capt. Elliott Springs	16
Lt. George Vaughn	13
Capt. David Putnam †	13
Maj. Reed Landis	12
Capt. Eugene Kindley	12

* Who served in the USAS at some point. For example, Raoul Lufberry scored most of his victories flying with the French, and George Vaughn scored seven of his 13 victories flying with the British; some American aces served only with British units and are not listed here.

Above: These pilots of Austro-Hungarian *Flik* 51J, photographed on August 1, 1918, accounted for 71 confirmed victories. From left, they are *Stabsfeldwebel* Stefan Fejes (16 victories), *Oblt.d.R.* Michael Dorcic, *Oblt.d.R.* Benno Fiala *Ritter* von Fernberg (Commanding Officer, 28 victories), *Oblt.* Franz Rudorfer (11 victories), and *Feldwebel* Eugen Bönsch (16 victories). European societies of the time were far more class conscious than at present, and this was especially the case with the Austro-Hungarian Empire as reflected in this photo; the three officers are at center with the NCOs flanking them at a distance.

Appendix 2: Frontbestand Inventory of German Biplane Fighters at the Front

Manufacturer and Type		1914			1915						1916						1917						1918			
		31 Aug	31 Oct	31 Dec	28 Feb	30 Apr	30 Jun	31 Aug	31 Oct	31 Dec	28 Feb	30 Apr	30 Jun	31 Aug	31 Oct	31 Dec	28 Feb	30 Apr	30 Jun	31 Aug	31 Oct	31 Dec	28 Feb	30 Apr	30 Jun	31 Aug
AEG	D.I																			1						
Albatros	D.I													1	50	39	28	20	17	12	9	8	5	6	1	3
	D.II													1	28	214	150	107	72	44	11	6	2	2	2	2
	D.III															13	137	327	303	385	446	423	357	174	82	52
	D.V																		216	424	526	513	250	131	91	20
	D.Va																				53	186	475	928	604	307
	D Lf																	2								
Aviatik	D.I																	15								
	D.III																	4								
DFW	D.II																		1							
Euler	D.II														2							11	16	13	6	1
Fokker	D.I													10	74	4	5	5	6	2	1					
	D.II													10	49	68	49	33	26	10			6			
	D.III													7	6	34	7			1		4		11	11	10
	D.IV													2		2					4					
	D.V																3	6	4	4	4	12	25	19	6	3
	D.VI																								21	27
	D.VII																							19	407	828
Halberstadt	D.I														6		12	16	14	1	2					
	D.II												6	5		55	5	2	14	16	3	2	1		1	1
	D.III										1		2	20	32	17	12	9	8	6	7	5	1	2	2	2
	D.V														17	32	39	11	6	9	5	3	2			
LVG	D.I																61	47	18							
	D.II																1									
Pfalz	D.III																			3	145	276	182	13	13	3
	D.IIIa																					114	261	433	289	166
	D.VIII																								14	19
	D.XII																								5	168
Roland	D.I														1	7	12	2	1							
	D.II																22	97	41	10	2					
	D.IIa																		128	102	6	3				
	D.III																						9	14		
	D.VIa																								38	58
	D.VIb																								17	12
SSW	D.I																	2								
	D.III																							4	4	6
	D.IV																									3
Total:											1		8	62	265	485	562	686	875	1030	1224	1566	1592	1769	1614	1691

Appendix 3: Frontbestand Inventory of German Triplane Fighters at the Front

Manufacturer and Type		1914			1915						1916						1917						1918			
		31 Aug	31 Oct	31 Dec	28 Feb	30 Apr	30 Jun	31 Aug	31 Oct	31 Dec	28 Feb	30 Apr	30 Jun	31 Aug	31 Oct	31 Dec	28 Feb	30 Apr	30 Jun	31 Aug	31 Oct	31 Dec	28 Feb	30 Apr	30 Jun	31 Aug
Fokker	Dr.I																			2	17	35	143	171	118	65
Pfalz	Dr.I																							9	7	1
Total:																				2	17	35	143	180	125	66

Selected Bibliography

Books

Bronnenkant, Lance, PhD., *The Blue Max Airmen*, Vols. 1–6, Aeronaut Books, 2012–2015.

Davilla, Dr. James J., and Soltan, Arthur M., *French Aircraft of the First World War*, Flying Machines Press, 1997.

Franks, N.L.R., Bailey, F.W, and Guest, R., *Above the Lines*, Grub Street, 1993.

Franks, N.L.R., Bailey, F.W, and Duiven, R., *The Jasta Pilots*, Grub Street, 1996.

Gray, Peter, and Thetford, Owen, *German Aircraft of the First World War*, second revised edition, New York: Doubleday & Company, Inc., 1970.

Green, William, and Swanborough, Gordon, *The Complete Book of Fighters*, Smithmark Publishers, 1994.

Grosz, P.M., *Junkers D.I*, Albatros Productions, Ltd., 1992.

Grosz, P.M., and Volker Koos, *Fokker Flugzeugwerke in Deutschland 1912–1921*, HEEL Verlag, GmbH, 2004.

Hallion, Richard P., *Rise of the Fighter Aircraft 1914–1918*, The Nautical & Aviation Publishing Company of America, 1984.

Herris, Jack, *The German Fighter Competitions of 1918*, Aeronaut Books, 2013.

Herris, Jack, *Pfalz Aircraft of WWI*, Aeronaut Books, 2012.

Herris, Jack, *SPAD Two-Seat Fighters of World War I*, Flying Machines Press, 2005.

Herris, Jack, *SSW Aircraft of WWI*, Aeronaut Books, 2014.

Lange, Bruno, *Das Buch der Deutschen Luftfahrtechnik*, Verlag Dieter Hoffman, Mainz, 1970.

Rogers, Les, *British Aviation Squadron Markings of World War I*, Schiffer, 2001.

Scott, Josef, *Fokker Eindecker Compendium 1 & 2*, Albatros Productions, 2012.

Articles

Grosz, Peter M., "Frontbestand" *WW1 Aero* No.107, Dec. 1985, p.60–66.

Grosz, Peter M., "Frontbestand" *WW1 Aero* No.108, Feb. 1986, p.66–69.

Above: A group of surrendered Fokker D.VII fighters at the USAS center in Romorantin, France, after the war. More than 160 of these fighters were shipped back to the U.S.A.

Selected Index

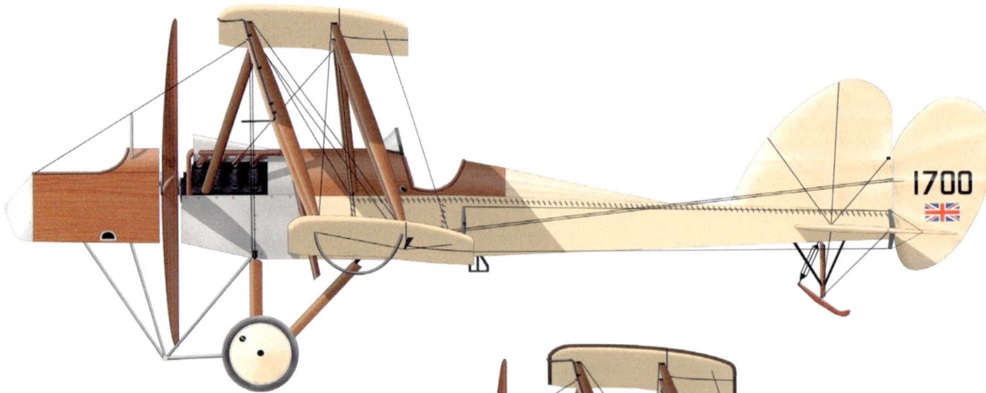

BE.9 1700 as tested at Farnborough, August 1915.

Note: The extent of the PC10 coloring on the BE.9 is speculative; the reference photo is dark and the nacelle and forward panels may have been left natural finish as shown above.

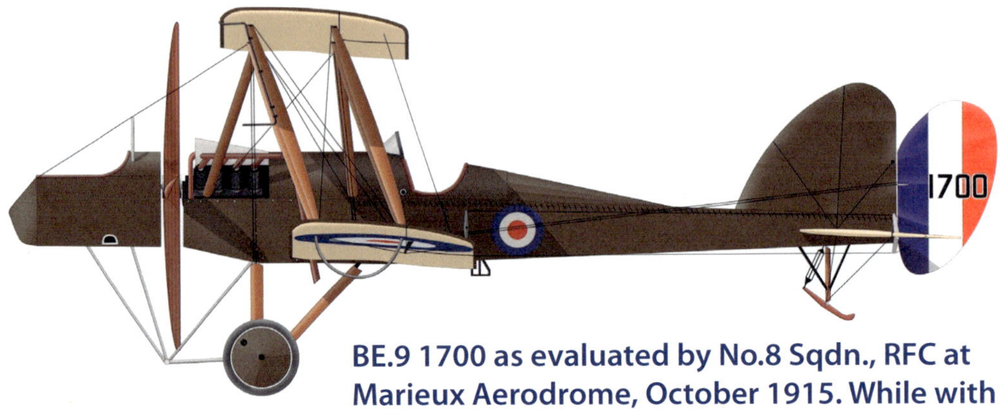

BE.9 1700 as evaluated by No.8 Sqdn., RFC at Marieux Aerodrome, October 1915. While with No.8 Sqdn. It encountered an *Eindecker* on 13 October but neither aircraft prevailed and the BE.9 was withdrawn by 11 December.

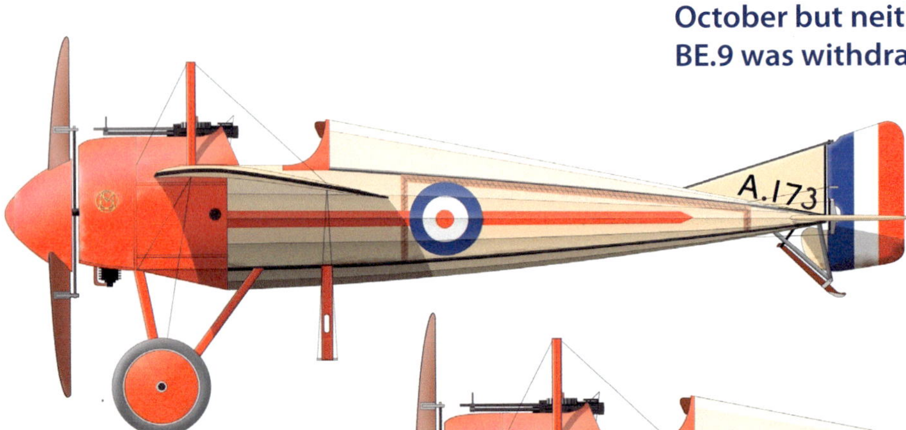

Morane-Saulnier N A.173 of 2/Lt. B.M. Wainwright, No.60 Sqdn., RFC, August 1916. Wainwright was brought down in this aircraft & made PoW on 28 August 1916.

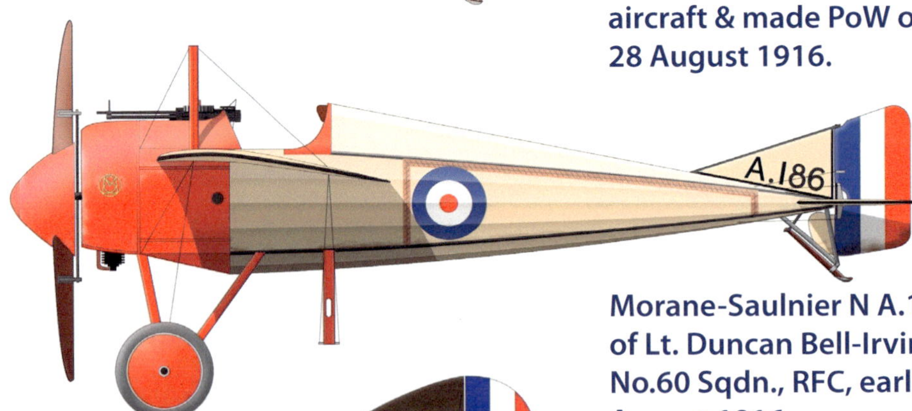

Morane-Saulnier N A.186 of Lt. Duncan Bell-Irving, No.60 Sqdn., RFC, early August 1916.

Bristol M.1C C4912 of No.150 Sqdn, RAF, Macedonia, 1918.

LUSAC-11 SC-42131, the fourth production aircraft, arrived in France for evaluation by the USAS on 17 October and first flew on 4 November 1918.

Afterword

Above: Of the millions of men who fought in the Great War, only one is remembered by most people today; the Red Baron, Manfred von Richthofen.

Above: The Ball Memorial on the grounds of Nottingham Castle shows the high regard for the fighter ace during a war that claimed millions of lives.

The story of the genesis of fighter aviation is a modern story of technological development seized upon by men eager to challenge the sky, something humanity had always dreamed about. And that makes it an intensely human story as well. These young men created a new kind of warfare and became a new type of warrior, something unique in history but grounded in the warrior tradition of courage, duty, and commitment to a goal greater than themselves.

The pilots flew fragile, unreliable aircraft against a determined enemy while exposed to severe cold and lack of oxygen, typically without parachutes to enable escape in case of disaster, which was close to all of them. And they were young, often teenagers.

Nations were quick to recognize their heroic example, especially in the face of the appalling and anonymous mass carnage in the trenches. At a time when the death and destruction surpassed any horror previously experienced, the fighter pilots demonstrated there was still a place for individual skill and courage, and sometimes even for mercy and chivalry. This was a ray of hope in a world of chaos and insanity, and it is no coincidence that the only soldier who fought in the Great War that is remembered by a general audience today is the war's greatest fighter ace, Manfred von Richthofen, known to legend as the Red Baron.

If there was a ray of light due to fighter pilots, the darkness was pervasive and the reality was grim. If the pilots lived better than their comrades in the trenches, they still died terrible deaths at a young age. Many were in their teens, few were older than their mid-twenties. If their fragile aircraft were hit, they easily caught fire or collapsed into scraps of wood and canvas. The vast majority of fighter pilots

Above: German flyers received 150 Marks per month danger money, later changed to 5 Marks per day. This is their sardonic comment on their flight pay!

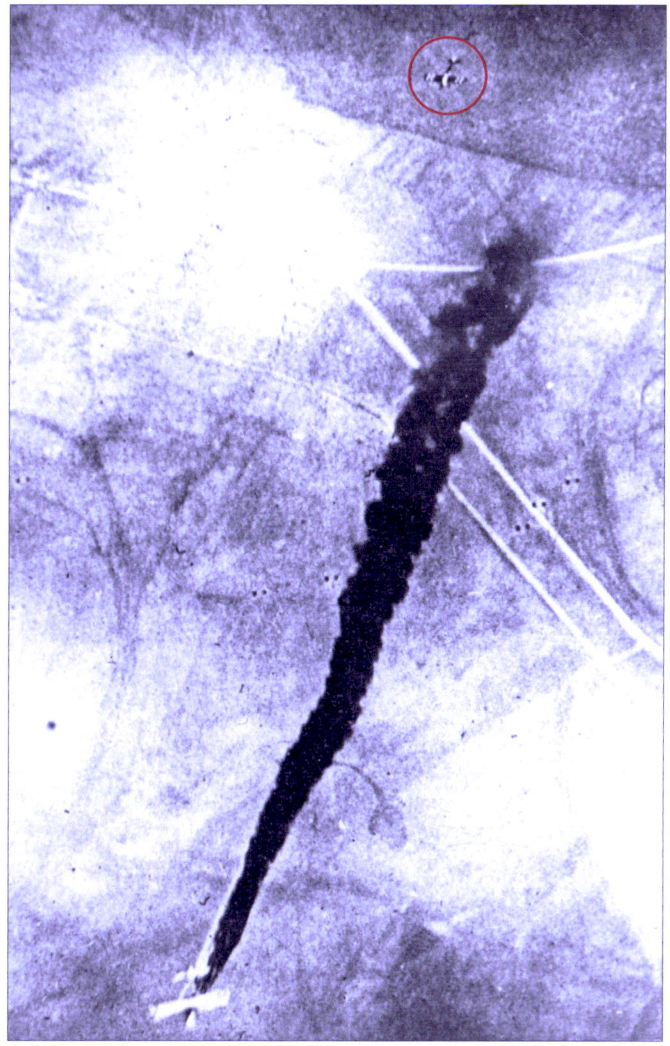

Above: An *Eindecker* (supposedly Boelcke), at top, scores and another Allied two-seater plummets into the abyss. These two photos illustrate the knightly vision's dark side.

never had a parachute, leaving them the option of riding a burning airplane to the ground or jumping to certain death. The average life expectancy of a pilot in combat at times dropped as low as three weeks, and many were lost in the first few days when their lack of air combat experience was fatal.

Flight training was primitive and there was very limited understanding of aeronautical principles. Many pilots perished in spins before spin recovery procedures were developed and generally taught to trainees. Aviation technology was primitive and effective communications and navigation were almost non-existent. Aircraft design was an art, not a science, and the only way to prove an aircraft design was for a test pilot to fly it, a process that resulted in great expenditure of test pilots.

Despite these challenges and a hostile operating environment, aviation developed into a critical component of military forces, and armies could not hope to win battles without at least rough parity in the air. Fighter planes and experienced fighter pilots became increasingly effective and essential to victory.

The pioneers of this new type of warfare leveraged their experience to become key commanders in the next great world war, in which air power was decisive. For the Allies this meant victory.

Fighter and attack aviation has now evolved into the most powerful facet of military airpower, replacing bombers as the primary instrument of airborne destruction. The technical sophistication and associated training required are difficult for non-specialists to imagine, but the human element remains. The courageous example of the first fighter pilots in their primitive and dangerous aircraft inspires us still.

Above: The *Pour le Mérite*, popularly known as 'the Blue Max', was the highest Prussian award and was only bestowed on officers. Because Prussia was the leading state in Imperial Germany, the *Pour le Mérite* served as the highest award of Imperial Germany and a number of aces were killed in action trying to qualify for the award, a feat that eventually required a minimum of 20 victories. Of the 81 awards to German aviators, 62 were to fighter aces.

Right: The highest British award for valour is the Victoria Cross. Of the 19 Victoria Crosses awarded to British aviators during the war, 13 were to fighter pilots, just slightly less than the proportion of German fighter pilots who were awarded the *Pour le Mérite*.

Left: In contrast to both Germany and Great Britain, only 3 of the 8 awards of the Medal of Honor to American aviators during WWI were to fighter pilots.

Above: The Order of St. George was the highest award officers in Imperial Russia could achieve.

Made in the USA
Monee, IL
12 September 2026

62976612R00169